T5-ASN-613

HOW SAFE IS SAFE ENOUGH?

How Safe is Safe Enough?

Leadership, Safety and Risk Management

COLONEL GREG ALSTON

ASHGATE

658.47
A46h

© Colonel Greg Alston 2003

All rights reserved. No part of this publication may be reproduced, stored in a retrieval system, or transmitted in any form or by any means, electronic, mechanical, photocopying, recording or otherwise without the prior permission of the publisher.

Colonel Greg Alston has asserted his right under the Copyright, Designs and Patents Act, 1988, to be identified as author of this work.

Published by
Ashgate Publishing Limited
Gower House
Croft Road
Aldershot
Hampshire GU11 3HR
England

Ashgate Publishing Company
Suite 420
101 Cherry Street
Burlington, VT 05401-4405
USA

Ashgate website: http://www.ashgate.com

British Library Cataloguing in Publication Data
Alston, Greg
   How safe is safe enough? : leadership, safety and risk
   management
   1.Risk management 2.Leadership 3.Industrial safety
   I.Title
   658.4

Library of Congress Cataloging-in-Publication Data
Alston, Greg, 1951-
   How safe is safe enough? : leadership, safety and risk management / Greg Alston.
      p. cm.
   Includes bibliographical references and index.
   ISBN 0-7546-3891-X
   1. Risk management. 2. Risk assessment. 3. Leadership. I. Title.

   HD61.A45 2003
   658.4'7--dc21

                                                            2003056046

ISBN 0 7546 3891 X

MI~

Printed and bound in Great Britain by MPG Books Ltd, Bodmin, Cornwall

Contents

University Libraries
Carnegie Mellon University
Pittsburgh PA 15213-3890

Disclaimer

The views presented in this book are those of the author and do not necessarily represent the views of the Department of Defense, the United States Air Force, or any component of these Agencies.

About the Author

Col Greg Alston is a 28-year fighter pilot in the U.S. Air Force. He has flown AT-38s, F-4Cs, F-4Ds, F-16As, F-16Cs, F-16CJs and F-117A Stealth Fighters.

Col Alston's safety experience began in 1991 as the Chief of Flight Safety Programs in the Pentagon, in Washington, D.C. In 1995 he was assigned as the Chief of Safety for Holloman AFB in Alamogordo, New Mexico, with a base population of over 6,000 personnel. While there, he won the National Safety Council Award of Honor for safety, became a squadron commander, and flew the F-117 Stealth Fighter. In 1997 he relocated to the Air Force Safety Center at Kirtland AFB in Albuquerque, New Mexico as the Deputy Chief of Safety Plans, Programs and Policies Division. Col Alston was re-assigned in 1999 to become the Director of Safety for the Air Combat Command at Langley AFB, Virginia, overseeing safety activities for assigned and gained units totaling 170,000 people and 1700 aircraft. There, he led the safety program to all-time lows in mishap rates for two years straight, records that still stand today. In 2001 he became the Deputy Chief of Safety for the Air Force at the Pentagon.

Col Alston is an Adjunct Assistant Professor with Embry-Riddle Aeronautical University, and has taught undergraduate and graduate level courses since 1988. His courses include Aircraft Accident Investigation, System Safety, Aircraft Structural Safety, Safety Program Management, Advanced Aerodynamics, Meteorology, Air Carrier Operations, Organizational Behavior, and Strategic Management. He is on the advisory board for Embry-Riddle's Center for Aerospace Safety Education.

Since beginning his safety career in 1991, Col Alston has become known as an expert in risk management, and helped develop the personal risk management concept. He has written and published 26 safety articles, many of which have focused on managing personal risks. During his two years as Director of Safety for the Air Combat Command, he was responsible for the production of the command's monthly safety magazine, *The Combat Edge*. *The Combat Edge* focused on flight, ground, and weapons safety, and commonly zeroed in on risk management issues. During that time, the Air Combat Command experienced its lowest mishap rates in its history in flight and ground operations.

List of Figures

Foreword

A persistent question that all leaders must face, from CEOs to team chiefs, is 'How safe is safe enough?' A simple question, five words that have far reaching effects on every organization. While the question is static, the answer is dynamic, ever changing as conditions change. The question is so incredibly important, that failure to answer it properly can end a junior manager's career, get a CEO fired, or even destroy an entire organization. To answer the question appropriately, leaders at all levels must become familiar with the nine elements of organizational risk reduction found in Chapter 1, and strive for awareness of possible hazards and universal threats.

The universe is a dangerous place. Hazards and associated risks abound. During my years in safety, I have seen many organizations fail to manage risks properly, and thus fail to protect people, guard assets or preserve the environment. Safety is not easy, it is a full-time effort, and is equally important whether people are on the job or on personal time. If an organization is serious about mission success, it must take 'risk' seriously as well. Leaders need to be involved in the risk game at every turn, and understand the key elements (discussed throughout this book) that help them answer the question–*how safe is safe enough*–and to win against risk.

Winning the risk game is what safety is all about. As in operational success, risk management requires the best human faculties to achieve victory; talent of organizational players and commitment from top, middle, and lower leadership rule the day. I have watched over safety teams, and found that those with committed support of senior leaders tend to win. However, leaders at all levels play a critical role in determining the correct level of safety. Over the years, I have observed seven truths in the safety game that leaders and managers must embrace:

1. All accidents are avoidable.
2. Achieving 'zero' accidents is possible (though often blocked by the human factor; the fallible human condition).
3. Risk comes with all activity: the benefit of an activity must outweigh the associated risk.
4. Safety is integral to operational success.
5. While everyone is responsible for safety, leaders have ultimate responsibility.
6. Without accountability, no one is responsible for safety.
7. An organization is only as safe as the leader allows it to be.

If we subscribe to the belief that all accidents are avoidable, then the possibility of achieving 'zero' accidents exists. I believe zero is possible, but I am also aware that every activity bears a certain level of risk, and over time probabilities may play out (Chapter 4). In the complexities of organizational activities (all bearing risk), and considering the fallible human condition (discussed in detail throughout this book), it is unlikely we will reach and sustain a level of zero accidents any time soon. It is in this light, and at this time in our human evolution, that leaders must strive for an optimum 'safe enough' organization while pursuing 'zero' in the long term.

Previously, as Chief of Safety for two large, high-risk organizations, I had no choice but to engage risk head-on. I was lucky to have great organizational leaders who played the risk game to win. Safety success requires leader awareness of the above listed truths. Another 'winning' ingredient is the understanding of what roadblocks stand in the way of achieving 'zero' mishaps–after all, zero is the ultimate goal. In the most basic sense, the challenges for leaders on the road to 'zero' mishaps are:

1. Acts of God
2. Limited resources
3. Unavailable technology
4. The human factor.

Of these, the human factor is key to winning or losing the risk game, and by far the most challenging. If fiscal resources are not available, procedural controls adjust risk to a temporarily acceptable level. If technology is not yet obtainable, again procedures or limited technology are available to help reduce risk until adequate technology is developed. Acts of God are tough to avoid, but through study, education, resource commitment, and technology, we can, over time, predict and avoid most events in this category. Acts of God are the 'natural' random events occurring in the universe, such as a deer jumping out on the road at night striking a vehicle, or a lightning strike during a thunderstorm. Humans can intervene in such natural events (Chapter 4). In natural events such as 'deer' strikes, people erect fences, drive slower, select other routes, etc., to avoid wildlife on the roads. Human decisions regarding resources, technology, acts of God control an organization's destiny, and determine *how safe is safe enough*.

All leaders and managers must be aware that the human factor is timelessly complicated. We know for example, that humans make decisions on acquisition, design, rules, maintenance practices, operational procedures, organizational structure, personnel talent, mission, and so on–all subject to error. Human interactions preside over all our activities. Since the human condition is fallible (Chapters 7 and 10), and we make mistakes, and leaders are human, our decisions are sometimes less than optimal, allowing risk to sneak into our lives and organizations. However, leaders who familiarize themselves with the nine elements of organizational risk reduction found in Chapter 1 can safe-up operations under their span of control.

Central to winning the risk game, leaders make the ever-critical decisions regarding *risk avoidance and risk acceptance* in their ideal quest for zero mishaps. Achieving zero mishaps is quite difficult, and sometimes unrealistic depending on many factors that lead to irresolvable risk. Until we are able to reach and sustain zero mishaps, leaders must make an important decision on the question, '*How safe is safe enough?*' Those who make the wisest decisions on this critical question *win the risk game.*

Chapter 1

Managing Risk in an Uncertain World

'We are deeply involved in fighting the war on terrorism and I compliment each of you on the success in that endeavor. Protection of our assets so that we can continue this fight is absolutely imperative. I need your support in the trenches to make sure that we develop the right mindset about risk management and doing the job right. It's hard and it takes time but there is no better way to ensure that we as commanders protect our most valuable asset, the men and women of the U.S. Air Force.' General John Jumper, Chief of Staff of the U.S. Air Force, safety message to commanders, June 2002

It was a beautiful day. The deep blue sky and calm air promised serenity; the comforting warmth offered a sense of peace. I assumed it would be a routine day in the Pentagon, wishing more that I could be outside enjoying the tranquil summer morning. Too soon, the news broke of a terror attack on the World Trade Center in New York; the sight on the viewing screen was all consuming. My fixation on the burning towers, however, was abruptly interrupted. The Pentagon jolted with a loud bang as American Airlines Flight 77 slammed into the base of the wall at Corridor Four. I gazed from my 5th floor window to the opposing side of the building to see the huge orange fireball with billowing black smoke rising into the air. As the Pentagon alarm sounded, I evacuated with the crowd along the 'E' Ring hallway amidst a scrolling veil of smoke. I proceeded down the stairwell to the exit at Corridor Two. Only minutes after the attack, I stood in the south parking lot and watched the Pentagon burn, and I wondered how we failed to see it coming. A new order was about to take shape for America, and the world.

Countries in all regions discovered they could no longer overlook risk from outside their borders, and as sobering a thought, from within their borders. National risk management requires we go to the source of the threat. In the case of the attack on 11 September 2001, eliminating risk requires we destroy the terror infrastructure globally, to include elements within our own nations, and root out and destroy terrorists, one by one if necessary. National security and the well-being of civilization as a whole depends upon awareness and aggressive action by our leaders to eliminate such risks, and to ensure we are *safe enough*. This is one example of how leadership contributes to winning the risk game and enhances our survival. While terrorists present one kind of threat, individual and organizational risks come in many forms. Both apparent and subtle hazards exist that we cannot overlook, which pose different threats and require aggressive action, and a change in our approach toward risk.

This morning, 1 February 2003, I watched on television yet another catastrophic event in safety, the break-up of the Space Shuttle Columbia. The nation, and the world, was in shock, observing an international accident of horrific proportion. The glowing streak resembled a sad, surreal tear for humanity, streaming from space over the Texas sky, settling upon its origin Earth. While it was a somber sight that touched our very souls, it was also one that will raise serious questions on safety. The ensuing investigation will answer the question, 'Is the NASA space program safe enough?' A parallel question faces all leaders of all organizations, 'How safe is safe enough?' It is an ever-present question that must be revisited as organizational conditions change, as new technologies evolve, and as global perceptions dictate.

As a fighter pilot in the U.S. Air Force, I was taught to know the enemy. I had to be aware of the weapons they will use to kill me (both from the air and from the ground). I studied their tactics, their numbers and any other information that could help me to win, or at least to help me stay alive in order to fight again another day. I considered the 'threat' in every aspect of mission planning to ensure success and survival–to ensure I was *safe enough*, yet still accomplish the mission. In essence, my *awareness* was heightened as I searched for hazards along my path, considered options to eliminate or reduce those hazards, and selected the option with the least risk that still allowed mission accomplishment. Fighter pilot thinking is not unlike the processes required by organizational leaders to accomplish their missions with the least amount of risk, by ensuring their organizations are *safe enough*.

Perhaps the recent historic events of '9/11' and the Space Shuttle Columbia accident will awaken leaders globally to take a similarly active role as fighter pilots into building a safety culture within their respective organizations. With these recent catastrophic events, America and the free world lost a fragile innocence, but so has each organization and each individual. We perceive our universe differently now, though it remains the same as it always was. Our indifference to risk has changed. Where once we may have inconsiderately strove for a blissful coexistence with universal characters, we must now strive for awareness of threats that share our place in time. The threats, hazards and risks have always existed, yet people have often chosen to ignore them, or lacked the initiative to find the unknown unknowns–a losing strategy.

Awareness, motivation, and commitment from leaders at all levels factor into winning the risk game. We cannot ignore reality, organizational leaders have no choice–they must coexist with risk. To ensure they are *safe enough*, however, organizations can choose to better navigate through the probabilities and severities of the universal risks they face. Awareness plays a key role for leaders in risk reduction, as they address four basic questions:

1. What are the hazards associated with our activity?
2. Which hazards can we eliminate or control?

3. Does the benefit of the activity outweigh the leftover risks?
4. Can we live with the result (consequence) if the probabilities play out in the worst way?

Once awareness is achieved, motivation is simple: mishap prevention is a profit multiplier. The financial bottom line is the driving force for risk management, but seeking out threats and reducing risks takes corporate commitment. An organizational commitment to such *threat assessment* begins the risk reduction process:

1. Assess all known threats.
2. Seek out unknown threats.
3. Consider options to eliminate or mitigate risks.
4. Eliminate risks where possible.
5. Control the risks you cannot eliminate.
6. Monitor and reassess.

In their quest to be *safe enough*, the above process helps management to win the risk game. However, leaders at all levels, not only senior executives, need awareness of the general concept of risk, and the indisputable fact that accepting risk is a choice.

The term 'risk' derives from the early Italian *risicare*, which means, 'to dare'. In this sense, risk is a choice rather than a fate (Bernstein, 1996). The actions we dare to take, or not take, are what the risk game is all about. Hazards abound, and each pose risks that are assessed comparing their probability of occurrence and the severity of the result should the probability play out. Taking no action to reduce risk is also a choice, daring natural events to simply not happen. Non-action may be a leader's choice if the assessment is, 'We are safe enough'. Such a choice must weigh the organizational efforts to eliminate or reduce risks compared to all the variables, such as available resources, mission, short-term and strategic goals, and the possible legal perils of non-action to a known risk. When combating risk, it is generally accepted that action is a much better choice than no action, where we can intervene in the probability-severity mix (Chapter 4).

Risks present *problems* in personal lives, organizations and entire nations, problems we must not ignore. As we coexist with risk, we must lead with the proper choices to minimize our exposure. We need to gain awareness of hazards, and then expend energy and effort to eliminate or control the risks; otherwise we are not *safe enough*.

'Never Walk Past a Problem.' General Ralph E. Eberhart, Commander, Air Combat Command

General Eberhart made the above quote to a Commanders Conference at Langley Air Force Base in Virginia in 1999. Having a zeal for safety, his comments to leaders who could make a difference were appreciated. Too often leaders appear aware of a problem, individually and corporately, yet fail to act. These failures occur for a variety of reasons–busy schedules, preoccupation, limited resources, inadequate energy, or in the most troubling cases, negligence. Too often, the first indication of negligence occurs when the accident investigation is underway. Such leadership renders an organization not *safe enough*. To win the risk game: achieve awareness first, of risks and responsibilities to counter them, and then confront the *problems* facing our personal and organizational worlds.

We live in a multi-planed universe, where hazards abound that we must not walk past. We measure our well-being in many ways: health, job, organization, group, relationship, family, image and quality of life–these often measure how we 'stack up' in life. Routine attacks by hazards that assault our quality of life 'determinants' can diminish our lives, cause havoc on corporate operations, and ruin relationships; yet, we can eliminate or reduce many of those hazards with awareness and personal effort. Along with our new awakening to threats and associated risks, we have a duty to ourselves, stakeholders, and associations to conduct our lives differently; with our eyes open and our spirits ready to act to be *safe enough*, both for personal well-being and corporate survival.

In our daily activities, we normally do not come face to face with terrorists, but we face equally destructive risks that must be managed for optimum performance and quality of life. Individuals make up organizations and have similar associated risks, but group goals and market shares have risk as well. Our awareness mindset can serve as a foundation to build a new mindful order for our group, division, or corporation ultimately to win the risk game. Without a concerted effort to thwart the risks, losing is the result.

Losing the Risk Game

Safety occupies extreme importance to every organization and individual, and its achievement occurs when proper risk management becomes a part of the culture of daily operations. If we are not *safe enough* and lack risk awareness, or ignore the need for safeguards, it remains just a matter of time until an accident occurs. The primary way organizations gain motivation toward a solid safety program rests in considering the stakes. Inadequate risk management costs a great deal. Generally speaking, the best argument for a strong risk management program is the cost of not having one. Cost measurement happens in several ways (Chapter 3), sometimes involving human injury or death, damaged corporate image, or lost potential. Routinely, tangible and intangible costs reduce to a dollar amount. When a person dies unnecessarily, the organization faces direct and indirect dollar tangibles such as insurance deductibles, training costs, and lost production. The intangibles include the obvious emotional costs of grief, sadness, and decreased morale that affect the organization. The corporate distractions that accompany grief and low morale impact job performance and, in the end, production goals. Thus, winning the risk game is imperative to organizational success.

Winning the Risk Game

Most people find fighting hazards a difficult, full-time job. In addition, seeking out and reducing hazards can challenge existing resources of time and money. Why then, should organizational leaders incorporate a safety game plan that identifies and eliminates or controls risks? Because ultimately, 'safety' functions as an operational enhancer and a profit multiplier. A well-run safety program offers positive dividends to any organization and can achieve *'safe enough'* status, and possibly reach zero accidents. Organizations can win the risk game, but they must play as a team, with guidance from the corporate top and commitment at every leadership level down to the individuals performing actions.

From 1999-2001, the U.S. Air Force's Air Combat Command enjoyed two great commanders, General Ralph Eberhart and General John Jumper. They both provided insights and an unwavering commitment to safety. They established a foundation that gave the charter to fight and win the risk war. Fortunately, these top leaders influenced an unparalleled success for two years, achieving the lowest mishap rates in the command's history in both flight and ground safety; records that still stand today. Most importantly, lives were saved and valuable assets preserved that now help prosecute the war on terrorism. Leadership's active role made this success possible, but it also required a team effort.

Success required a 24 hours a day, 7 days per week effort from all members in the command to beat the odds against the numerous hazards that the Air Combat Command faced organizationally. While difficult, specific understanding of safety principles and risk management decisions helped leaders, managers, and individuals win the risk game in high-risk environments.

Managers in particular must put safety into perspective, and must make rational decisions about where safety can help meet the objectives of the organization. From an organizational perspective, safety is a method of conserving all forms of resources, including controlling costs. Safety allows the organization to pursue its production objectives without harm to human life or damage to equipment. Safety helps management achieve objectives with the least risk (*Flight Safety Digest*, Vol.13, 1994).

Managers must be aware of available hazard-reduction tools (expanded upon in other chapters) as well as the benefits of solid safety and risk management programs similar to those that helped the Air Combat Command. The same concepts that helped the generals *win the risk game* can help top executives, middle-managers, group leaders, line supervisors and individuals gain insight into risk in the broad sense, and help leaders answer the question, '*How safe is safe enough?*' Some 'key' leadership behaviors provide essential 'rational elements' of risk reduction that help leaders along the way. Importantly, these apply not only to CEOs, but also to all people in positions of leadership, from CEOs to middle managers to line supervisors. Essentially, all organizations can benefit by adhering to the same basic tactics to win the risk game, and ensure their organizations are *safe enough*. These are laid out in *nine elements of organizational risk reduction*–and become a risk management guide for leaders.

Nine Elements of Organizational Risk Reduction

Leaders must:

1. *Lead* the risk game.
2. Know the *costs of losing* the risk game.
3. Comprehend *universal probabilities*, and the effects of *human intervention*.
4. Understand basic principles of *risk management*.
5. Understand the basics and appreciate the value of the *system safety* process.
6. Be familiar with elements of *organizational risk*.
7. Appreciate the value of *personal risk management*.
8. Get involved in the organization's *safety program*.
9. Be open to positive *change*.

These key elements make up the essential tools organizational leaders need to beat risk and answer the safety question. The following chapters discuss each element in detail, and offer insights to managers at all levels on how to lead and win the risk game. In the end, or perhaps I should say the beginning, leaders need to be involved in safety and risk management, and continually answer the question, '*How safe is safe enough?*'

Chapter 2

Leading the Risk Game

'A company may in many ways be likened to a human body. It has a brain and a nerve centre which controls what it does. ....directors and managers....control what it does. ...the state of mind of these managers is the state of mind of the company and is treated by the law as such.' Lord Denning

Leaders show the way. They guide and persuade people toward a preferred course of action. Not all leaders are Chief Executive Officers (CEO's) or senior executives. Managers and supervisors at every level of an organization provide leadership. While their span of control differs, their function as leaders remains constant—accomplish the mission. A CEO directs executive actions to achieve strategic goals; a line supervisor leads a group of workers toward a task or daily quota. As part of the 'whole-team' effort, the daily quota contributes to the strategic goal. The interconnections represent an important corporate effort. Equally, in their sense of responsibility to show the way, organizational leaders are faced with an ever important and persistent question, 'How safe is safe enough?'

Safety must span all corporate levels because it provides the common thread that ties all activities together to enhance the end game. Safety acts as a profit multiplier when leaders focus on risk reduction. This can occur at every level, from strategic planning to the production line to the loading docks. However, the corporate effort to be 'safe enough' begins at the highest level. Senior executives lead the risk management effort from the top to set the tone. In concert with senior direction, mid-level managers must also lead risk management efforts in their areas; all leaders must show the way, and must all answer the safety question. When all is said and done, organizational leaders at every level are accountable and responsible for safety. Found in that truth, the responsibility is dual tracked:

1. A *fiscal* responsibility to enhance the bottom-line through sound safety practices and a solid risk management process.
2. A *moral* responsibility to organizational members, their families, stakeholders, and society in general to protect life, guard corporate assets, and preserve the environment.

Safety is placed squarely on the shoulders of leaders and managers, but must begin with top management. However, a leader cannot win the risk game alone; winning takes a team effort, buy-in, from the organization at large.

Achieving Buy-in

The CEO determines the importance of safety. When the CEO is involved in the safety program, the organization transforms into a safety-minded force with tentacles spanning the far reaches of the organization. Without buy-in, however, an organization is not *safe enough*, in spite of CEO rhetoric. Risk management at a corporate division depends upon buy-in from the division chief, who gets his direction from the executive level. The supervisor of the loading dock must then embrace the division chief's vision on safety and overall risk management. Corporate buy-in for risk management within any organization begins with the leader, who issues corporate vision that filters down throughout the leadership force. However, total buy-in by the corporate body depends on several factors. Leaders accomplish five tasks to achieve *buy-in*:

1. Hold people accountable for negligence regarding safety.
2. Provide incentives, or awards, for positive safety performance.
3. Mentor corporate members on safety and risk management.
4. Create a participative safety process for all members.
5. Be a champion of safety.

These tasks are important to merge leaders and organizational members with risk reduction efforts, and to enhance transformation to a strong safety culture (Chapter 7). Notably, as logical thought processes take place, leaders can achieve *organizational efficiency* as a by-product beyond safety when they lead the risk game. A review of the five leadership safety tasks is worthwhile.

Accountability

Leaders must hold people accountable for their actions. After all, governments will hold organizations accountable for safety regulatory violations, leading to fines and bad press, thus affecting the bottom-line. Organizations must require strict adherence to safety policies by its members, both on and off the job, and at every level from senior executives to line workers. Allowing one person a 'free pass' for a willful wrongdoing encourages others to expect similar tolerance–promoting an *unsafe* environment. This truth applies to all activities within an organization, including safety and risk acceptance, whether on or off the job. Leaders must note that organizational members who accept unnecessary risks 'off the job' (Chapter 8) also risk the corporate bottom-line–if people are injured on a weekend, the organization is diminished nonetheless. Therefore, some off-duty behaviors are subject to corporate accountability along with on the job violations. Permitting people to ignore risks, means they often 'will' for expedience or ease. When employees ignore risks the inevitable result allows probabilities to play out–an accident will occur. One can enforce accountability in several ways:

- Termination
- Fines
- Reduction in pay-grade
- Probation, with training
- Loss of privileges
- Denied bonuses
- Incorporation of safety objectives into performance appraisals.

These are punishments. Termination eliminates a future problem and enforces the importance of safety to remaining members. Fines, reduction in pay and probation are punishment tools that allow individuals to remain in the organization. Notations in a member's performance appraisal may actually be positive for sound safety performance, but presents a threat of permanent documentation for violations–a punishment.

Punishment is the attempt to eliminate or weaken undesirable behavior. It is used in two ways (Nelson and Quick, 2000). One way to punish a person is through the application of negative consequences following an undesirable behavior. The other way to punish is to withhold positive consequences, such as loss of an annual bonus. Problems can occur with punishments in the form of unintended yet *harmful* consequences. The experience of punishment may result in negative psychological and emotional behavior or poor performance. People may feel angry, hostile, depressed, or despondent. Therefore, leaders must use punishment carefully, and make sure it is justified.

Managers must make the important distinction between an honest human mistake and an intentional disregard of safety guidelines. Human error or other human factors contribute to approximately 90 percent of all accidents: 70-90 percent of system failures (Bahr, 1997), 78 percent of aircraft crashes (Krause, 1996), and over 90 percent of automobile accidents (Green and Senders, 1997). At times, human factors such as optical illusions, spatial disorientation, and perception are not necessarily anyone's fault; they are simply human frailties that we try to avoid with training and/or technology. However, complacency, inattention to detail, and even fatigue could be determined as negligent in the cause of an accident.

Leaders must demonstrate a critical component of accountability–determining the root cause of erroneous behavior. If an employee has an accident, it may very well be the supervisor's fault for not providing adequate training or oversight. In general, people should be held accountable for the following:

- Willful violations of safety guidelines
- Willful disregard for company policies
- Negligence
- Complacency.

Accountability is an important tool to achieve group buy-in for safety. It enforces standards, strengthens corporate policy, and signals the importance placed on safety and risk management. Accountability, however, indicates there has been an organizational failure, and that someone was held to account for that breakdown. While accountability is a critically important instrument, other tools in the form of *incentives* are available that attempt to intervene human failures by rewarding prevention efforts.

Safety Incentives

Rewarding good safety practices is positive reinforcement. *Operant conditioning* is the process of modifying behavior through the use of positive or negative consequences following specific behaviors. Positive reinforcement is used to enhance desirable behavior, while punishment is used to diminish undesirable behavior. The application of reinforcement theory is central to the design and administration of organizational reward systems (Nelson and Quick, 2000). Considering there are negative effects associated with punishment, positive reinforcement is often the most desirable motivation system for an organization. Thorndike's *law of effect* states that behaviors followed by positive consequences are more likely to recur (Chaplin and Krawiec, 1960). Thus, rewarding safety performance, such as a team award for 30 days on the production line without a lost workday, will produce further safe operations.

The types of rewards can vary, but all must be publicized. Safety awards contribute to mishap prevention because others see the importance to the organization and gain safety awareness. There are different types of safety awards, depending on the safety goal. Some are group awards, while others are individual. If a production line has no injuries over a specified period, reward all employees. If an individual prevents an accident, or devises a great safety initiative, an individual award is warranted. Some possible award types are:

Non-monetary

- Employee of the month plaque
- Production plant of the quarter/year award
- Company-sponsored picnic on Friday afternoon (with pay)
- Gifts, with a safety message (flashlight, coffee cup w/lid, non-breakable sunglasses, etc.)
- Write-up in company newsletter
- Annotation in annual performance report.

Monetary

- Performance bonus
- Time off work with pay
- Promotion opportunities
- Travel awards
- Dinner for two.

Awards can be innovative, but must fit the positive safety act, and must reward the right people. If someone discovers bare electrical wires, takes the effort to block the hazard, and then promptly reports it to a supervisor, the reward is simple: his choice in a safety gift–coffee cup with a protective lid. If a person took quick action to put out a storage room fire, saving the entire plant from destruction, then a travel award (weekend away), promotion opportunity, or monetary award is in order. Advertise the award presentation to spread the word, and to enhance the overall safety mentoring efforts for future leaders.

Mentoring

'Seek out opportunities to mentor.' General Ralph E. Eberhart, as Commander, Air Combat Command, given during Wing Commanders Conference, Langley Air Force Base, Virginia, 1999

In the above quote, General Eberhart referred to our responsibilities as leaders of today to teach and mold tomorrow's leaders. *Webster* defines a mentor as a wise, loyal advisor, teacher, or coach. We have gained experience and knowledge over the years, and are now successful. How wise would it be to take the secrets of our success to our next lives? As well, if we value our organization, we must transfer our winning tactics and wisdom to the next generation. Our wisdom is not limited to leadership, but has elements of basic survival. Our survival success is wide-ranging from career progression to staying alive, and centers on effective risk management, to include personal risk management. We as wise teachers must pass these winning ways to future leaders.

Mentoring is a work relationship that encourages development and career enhancement for people moving through the career cycle. Mentor relationships typically go through four phases: initiation, cultivation, separation, and redefinition (Nelson and Quick, 2000). Throughout these stages, there is a side benefit of mentoring: the opportunity to mentor new employees can keep senior workers motivated and involved in the organization. While mentoring, it is vitally important to add the topic of safety and risk management into the typical career progression theme. Safety is everyone's business, but new members may not realize it unless leaders teach them. As stated before, leaders show the way in safety, and mentoring is an important avenue to make it happen while adding to a participative atmosphere.

Create a Participative Safety Process

An organization does not have a safety culture if some members are left out. Leaders must include everyone. The effects of organizational cultures are often debated among organizational behaviorists and researchers. Among the competing theories is the *strong culture perspective,* where organizations with 'strong' cultures perform better than others (Deal and Kennedy, 1982). A strong culture is an organizational culture with a consensus on the values that drive the company and with an intensity that is recognizable, even to outsiders. Thus, a strong safety culture is deeply held and widely shared (Nelson and Quick, 2000). In this light, leaders must include all organizational members.

Inclusion into the safety process begins during initial training. When a new employee enters an organization, safety training takes place. At that time the new member learns company policy, the role of the safety managers, and the responsibility of each person regarding safety and risk management. If a new person is convinced of the corporate enthusiasm toward risk reduction, then he or she is more apt to comply and buy-in is achieved; training will show members how they fit into the safety process. They must learn early on that hazard reporting is essential, and it is their responsibility. They must also be aware that they are held accountable for ignoring a known risk. Hazard identification and reporting must be a participative effort for all members. Along with training, other areas that encourage participation are:

- Safety awards, and punishments
- Group safety meetings
- Safety suggestion boxes
- Solicitation of ideas.

Enthusiastic participation is key to a culture. Leaders can obtain such a safety culture through inclusion of the corporate members, by building a participative environment toward safety and risk management. Members are also more apt to promote safety if the leaders are passionate toward winning the risk game.

Champion Safety

CEOs and other leaders achieve buy-in to organizational safety through the previously discussed topics: accountability, award systems, mentoring efforts, and a participative process. They have one more key step to effectively lead the risk game and ensure the organization is *safe enough*: be a corporate champion of safety. The leader of an organization must show the way, and set the example. Address topics on risk reduction and other safety ideas often, in nearly every setting. The CEO leads the effort to culturalize safety, but every mid-level manager must do the same. If leaders do not demonstrate an interest, then safety becomes less important among the masses. A good way to show interest is to mention the words 'safety' and 'risk' every day. Ultimately, leaders must address the critical question, 'Are we safe enough?'

Leaders can begin their day asking the question, 'Where will an accident happen today?' Follow that question with, 'What have I done to prevent it?' Leaders and managers can add other questions as desired to these in a leader's morning safety checklist. Each leader can build a personal specific checklist.

The Leaders' Morning Safety Checklist

- ❑ Where can an accident happen today?
- ❑ What have I done to prevent it?
- ❑ Are there new hazards facing my operation?
  - o Adverse weather?
  - o Are we beginning a new, unfamiliar operation?
  - o Are we introducing change?
- ❑ Is there organizational stress?
  - o Are there layoffs due to company right sizing?
  - o Are production plants in the process of relocating?
  - o Is the local community suffering economic depression?
- ❑ Call the director of safety for an update.
- ❑ (For CEOs) Call my Directors to discuss possible threats.

These are simple questions, but they achieve the goal of thinking 'safety'. They can ask the question, 'Today, are we safe enough?' Leaders should tailor the checklist to fit their needs, and can even have a checklist for each day of the week to avoid monotony and numbness to the process. In keeping with the notion that leaders should champion safety at every opportunity, checklist questions are an excellent tool to lead off a staff meeting, or to ask people during a site visit. Site visits are an important avenue for leaders to be champions for safety and risk reduction. The following questions are derived from the 10 safety questions in Fiscal Year 2002 that the U.S. Secretary of the Air Force and U.S. Air Force Chief of Staff used to incorporate safety into their base visits, and to help answer the question, 'Are we safe enough?'

Safety Questions to Ask During Site Visits

1. What are the most significant hazards your personnel are exposed to and what are you doing to control the risks? (Leaders/Line Supervisors)
2. How do you address personal safety concerns with your people? (Senior Leaders/Line Supervisors)
3. How do you hold your managers accountable for the safety of their personnel? (Senior Leaders/Directors)
4. How have you encouraged your subordinates to question a work practice they think is unsafe? (Senior Leaders/Line Supervisors)
5. Has your Risk Management program had an impact on your operations and if so, how? (Senior Leaders, Line Supervisors, Individuals)
6. Tell me how you communicate safety as an integral part of the daily mission. (Senior Leaders/Line Supervisors)
7. Tell me the best and worst safety training you ever had. (Individuals)
8. Tell me about the most hazardous task you accomplish and what safety precautions you are taking to control the risk. (Individuals)
9. Are you aware of any safety requirements that impede mission accomplishment? (Everyone)
10. How do you ensure your hazardous materials are properly stored and accounted for, and explosives materials properly sited? (Senior Leaders, Line Supervisors, Safety Managers, Individuals)

Note some questions are directed at senior leaders, some to middle management, and some to individuals. This approach gives safety a corporate structure and pursues the 'whole-team' concept.

If leaders (particularly top leaders) routinely ask safety questions, the entire organization gets the message that safety is important. Using checklists and question banks helps leaders to champion safety. As such, they show the way in fighting risk. It also helps them tie in the other four tasks to achieve safety buy-in: accountability, award systems, mentoring, and creating a participative process. Ultimately, such achievement is possible only when leaders lead the risk game, and when they strive to answer the question, 'How safe is safe enough?'

Chapter 3

Costs of Losing the Risk Game

'The driving force behind a safety program is the cost of not having one.'
Alston (2003)

Money is the universal language; corporate costs must be communicated to all organizational players to relay the magnitude of accidental losses. Profit-oriented organizations consider Return on Investment (ROI) in nearly every activity. Notwithstanding a leader's moral responsibility to protect lives, guard assets, and preserve the environment (a card well played by prominent safety-minded organizations), safety is a financial win-win proposition well worth the expended resources. Leaders must ensure their organizations are *safe enough* to achieve the mission.

In safety, out-of-pocket expenses are compared to the 'avoided' out-of-pocket expenses to determine ROI. The general rule is every dollar spent for safety must be equal to or less than the possible loss. A simple example is an activity where the possible loss would amount to $100,000. A CEO would not spend $200,000 on safety to avoid losing $100,000 in assets (unless moral obligations trumped fiscal concerns). The trick is to determine the actual corporate costs. Some direct costs are easy to tabulate, while intangible costs are often 'fuzzy' and may not be realized for years. An understanding of the tangible and intangible costs, as well as direct and indirect costs associated with accidental loss is essential for leadership buy-in to a risk management culture.

Direct Costs	Indirect Costs
Destroyed equipment	Loss of individual experience
Lost revenue	Lost time and overtime
Lost cargo	Low productivity from depression
Insurance deductibles	Disrupted schedules
Training new people	Restoration of order
Training future avoidance	Lost market share
Environmental clean-up/fines	Loss of use of equipment
OSHA fines/ Legal fees	Loss of business
Corporate liability	Corrective action (Direct or indirect)
Increased insurance premiums	

Figure 3.1 Tangible costs of an accident

Figure 3.1 gives a list of tangible costs of an accident when an organization loses the risk game. Direct costs are directly associated with the mishap. Indirect costs account for financial losses as a by-product of the mishap.

Figure 3.1 gives a good picture of why safety plays a vital role in organizational achievement and success. Comparing Figure 3.1 with intangible losses listed in Figure 3.2 shows an organization not only loses an object, or worse a person in an accident; pervasive effects range widely, ultimately threatening future activities. Intangible costs may have lingering effects for years, and harm the bottom line–productivity.

Company image	Organizational uncertainty
Union pressure	Human resource acquisitions
Damage to reputation	(New employees may be reluctant to
Morale	work for your company)
Lost workdays	Family members' worry & depression
Government scrutiny	Social distrust

Figure 3.2 Intangible costs of an accident

Some of the direct costs are recoverable, but indirect costs and intangible losses are not. There are two basic accident costs: (1) insured costs, generally including equipment losses, property damage, and personal liability; and (2) uninsured costs (*Flight Safety Digest*, Vol.13, No. 12, 1994). Insured costs–those covered by paying premiums to insurance companies–can be recovered to a greater or lesser extent. Uninsured costs cannot be recovered and sometimes triple the insured costs, and often linger for years to come. Organizations must strive to be *safe enough* to avoid such losses.

We recognize the importance of safety, and all leaders must attend to risk management. Risk management occupies the core of safety, and requires oversight by leadership as well as individual buy-in. However, others are as interested in risk reduction within an organization as managers and employees, people who suffer along with the corporate body.

Who suffers the most when a serious accident results in major costs? Stakeholders. Every organization has stakeholders who are interested in corporate efficiency and reduction of costs–these stakeholders have a vested interest in reducing risk, and include more than only organizational members. Such stakeholders help develop the company mission, and have an interest in safety. Various stakeholders in an organization are divided into two groups: inside stakeholders and outside stakeholders. Insiders are individuals or groups that are stockholders or employees. Outsiders are all the other individuals and groups the organization affects. They are listed in Figure 3.3 (Pearce and Robinson, 2000).

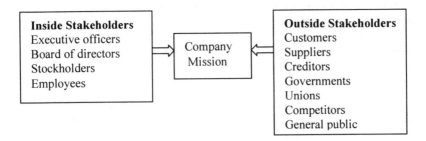

Figure 3.3 Stakeholder inputs to the company mission

Leaders must know the stakeholders who are at risk when a loss occurs, who are impacted financially, or in well-being. A typical list of stakeholders and their stakes are listed in Figure 3.4 (Pearce and Robinson, 2000).

Simply put, we all have a stake in safety, and want to ensure we are *safe enough*. When an accident occurs, society in general bears the burden, but the organization takes the immediate financial hit. This reaffirms that while we must coexist with risk, various processes can control risk by first identifying deficiencies and then correcting them. Senior management must ensure those processes are in place, and incorporate risk management into organizational objectives to conserve valuable resources for all stakeholders.

Management must put safety into perspective, and must make rational decisions about where safety can help meet the objectives of the organization. From an organizational perspective, safety is a method of conserving all forms of resources, including controlling costs (*Flight Safety Digest*, Vol.13, No. 12, 1994). Sometimes, however, accidents happen; leaders must make sure adequate insurance is available to lessen the risk to the organization. Insurance covers much of the cost, but does not cover everything, making safety vital to corporate and individual survival.

Insurance premiums are good investments. However, while most people think that their Employer's Liability Insurance will take care of all the costs if there is an accident, they must reconsider. It has been established that for every accident leading to serious injury there are approximately 10 minor injuries and 30 property damage accidents. For every $1.00 recovered by insurance, there is at least $11.00 of unrecoverable cost (Health and Safety Services, 2002). Studies conducted at Stanford University estimate the cost of accidents for users of commercial and industrial construction at $1.6 million annually; hidden costs were found to be 18 times higher (Bahr, 1997). Such disparity in recoverable costs lends criticality to a strong safety program, and factor into the question, 'How safe is safe enough?'

Just how bad can it get? While many organizations worry about lost workdays due to unnecessary injuries, in some cases an accident can cause indescribable disaster and grief. An accident in Bhopal, India in 1984 released

Stakeholder	Stakes
Stockholders	Profits, stock offerings, assets, stock transfers, elections
Creditors	Interest payments, return of principle, security of pledged assets, and certain manager prerogatives if promised
Employees	Economic, social, and psychological satisfaction, fringe benefits, adequate working conditions, bonuses
Customers	Service for the products, technical data for product use, spare parts, facilitation of credit, R&D for improvements
Suppliers	Continuing source of business, timely consummation of trade credit obligations, professional relationships
Governments	Taxes on income and property, interest in free and fair competition, discharge of legal obligations of businesspeople
Unions	Recognition as negotiating agent for employees, perpetuation of the union, participation in the organization
Competitors	Observation in the norms for competitive conduct established by society and industry
Local communities	Productive and healthful employment in the community, participation of company officials in community affairs and social events, support for cultural/charitable projects
The general public	The organization's participation in and contribution to society as a whole, assumption of fair proportion of the burden of government and society, fair price for products and advancement of state-of-the-art technology

Figure 3.4 Stakeholders and their stakes

methyl isocyanate and caused over 2,500 fatalities. In 1986, the NASA Space Shuttle *Challenger* disintegrated in flight in front of millions of viewers and killed seven astronauts, brought NASA to a standstill for two years, and cost the agency billions of dollars. A petroleum refinery blew up in Houston Texas in 1989, killing 23 workers, damaging property totaling $750 million, and spewing debris over an area of nine kilometers. In 1995, U.S. governmental statistics indicate that more than 350 chemical accidents occur each year that result in death, injury, or evacuation. Along with those figure, in 1991 and 1992 fifteen major petrochemical accidents destroyed more than $1 billion in property (Bahr, 1997). Such unthinkable accidents can indeed happen, sounding a siren for safety to reduce loss of resources and enhance social responsibility. Solid safety programs are essential.

The Chernobyl nuclear accident on April 26, 1986 perhaps shows the greatest magnitude of costs due to an accident. The cost to society as a whole was tremendous, and unacceptable. The litany of costs to society is staggering (Chernobyl, 2002):

1. Death rates are 30 percent higher for those in contaminated regions in the Ukraine compared to the rest of the country.
2. Birth rates in Belarus have fallen 50 percent.
3. Thyroid cancer, particularly among children, up 285 percent in Belarus.
4. About 7,000 in Russia alone who helped put out the fire and seal the reactor are believed to have died and 38 percent are recovering from some kind of disease.
5. Belarus, the most heavily affected country, spends 20 percent of its budget on dealing with Chernobyl's aftermath; Ukraine devotes four percent and Russia, one percent.
6. Contamination of Lake Kojanovskoe–downriver from Chernobyl and used by more than 30 million people–with 'radiation levels 60 times above European Union safety norms'.
7. Repair estimates for the disintegrating sarcophagus range from $1.28 to $2.3 billion.
8. 125,000 people alone have died 'from diseases related to the accident' according to Ukraine's Health Ministry.
9. Ivan Kenik, Belarus's Chernobyl minister, estimates the cost within the borders of Belarus for 'total damages from the Chernobyl catastrophe from 1986 to 2015' to be $235 billion.

In most cases, the organizations in the above examples were not *safe enough*. Strong safety programs can prevent such catastrophes. When the severity of the consequences of a risk is high, risk acceptance must be greatly scrutinized, and accepted risk held to the lowest possible level. The question must be revisited, 'Does the benefit of the activity outweigh the risk?' Can an organization stand the cost should an accident occur? Senior leaders must grapple with such questions on a strategic level, but also keep these questions in their crosscheck daily. Keeping abreast of organizational safety programs give the top leaders, and leaders at every level, the tools they need.

While safety programs prevent catastrophes, they also prevent individual lost workdays, a subject that seems small on the surface but can add up to a lot of cash. One program that promotes safety in the near and long term for the entire organization is system safety (Chapter 6). An organization can incorporate system safety into a system's life cycle, bringing with it management's commitment toward safety. Such a process can greatly reduce accidents, thus preserving assets and multiplying profits. According to the Occupational Safety and Health Administration (OSHA) (Bahr, 1997):

- Mobil Chemical Company cut its workers' compensation costs 70 percent, or more than $1.6 million, from 1981 to 1983.
- At Georgia Power's two power plant construction sites, the direct cost savings from accidents prevented at one site was $4.14 million at one plant and $.5 million at the other in 1986.
- Mobil Oil Company's Joliet, Illinois, refinery experienced a drop of 89 percent in its workers' compensation costs between 1987 and 1993.

OSHA's figures promote the claim: *The driving force behind a safety program is the cost of not having one.* Well-managed safety programs can indeed multiply corporate profits, and sometimes save the life of the organization. Some businesses have been forced to close because of accidents. It is estimated that about 250 million man-days of work, commonly referred to as 'lost workdays', are lost to accidents annually (Ferry, 1981). This equates to lost productivity and increased costs. If an average worker makes $15 per hour, he or she makes $120 per day in an 8-hour day. 250 million lost workdays in this example would amount to $30 billion annually due to accidents.

The fact is, an organization, and society in general, cannot afford accidents. In some cases, prosecutors will attempt to prove negligence on the part of corporate leaders if adequate safety and risk management tools are not in place. Beyond that is a primary motivator: winning the risk game multiplies profits. Organizational leaders must fight and win against risk, and properly answer the question, 'How safe is safe enough?'

Chapter 4

Universal Probabilities

'Allowing chance to control our lives renders us as pawns to universal ways.'
Alston (2003)

The Universe is an amazing place. The vastness alone is mind-boggling, not to mention the complexity of all the parts. The staggering size of our own Milky Way Galaxy includes 200 billion stars in its vortex, located in and between at least four major spiral arms and several short arm segments (Comins and Kaufmann, 2002). Consider there are literally billions of galaxies, some containing trillions of stars and innumerable solar systems and it becomes quite difficult to comprehend.

Countless activities occur at any one moment in our universe, perhaps some activities that are yet unknown to us. We cannot expect to know everything in a mere 30,000 years of existence for modern man. But out of this mysterious place, we know two things inherent with the universe; initially, with every random or intentional activity comes risk, and secondly, the timeless universe appears patient, seemingly in no hurry to carry out its journey. These two factors make risk a reality: risky action and enough time for the probability to come true.

Probabilities and consequences rule the very nature of the universe. Interestingly, this truth applies to building worlds and lives, as well as destroying planets and life. An important aspect to consider in our personal and organizational lives, *allowing chance to control our lives renders us pawns to universal ways*. Random dispersion predominates as the method of choice for the universe–stars race in groups of vortices by the billions and draw masses of matter into orbit. As the countless universal players draw into orbit, random collisions might build a larger planet, or might cause a degree of destruction on an unsuspecting world or other piece of matter. Our very own moon may have been formed from Earth's debris after an asteroid impact, know as the collision-ejection theory (Comins and Kaufmann, 2002). In pure randomness, an ice comet can collide with a mass and bring life-supporting and world-shaping water. However, that same comet can destroy life forms currently existing on the target, all by chance.

The universe, though still mysterious, contains a certainty of random actions, many of which we cannot control. The same can be true in our personal and corporate universes; we can fall prey to random and sometimes uncontrollable effects in our environments. Awareness, however, allows us to choose options and behaviors that affect probabilities and severities (consequences) in our lives. Aware choices can come from experience (age), training, education, and mentoring. Isn't it great when we can learn from others rather than repeating their failures or waiting until we are elderly to get smarter? However we do it, we must achieve awareness of universal ways.

Just like the bigger universe, indiscriminate acts can positively shape or completely destroy our individual worlds. Other players that share our world play a pivotal role. They might provide assistance or protection, or they might randomly attack our personal lives. Some accomplish this in obvious ways while others are quite subtle. Subtly, a virus can arbitrarily select a human body as its host, or E.Coli bacteria can invade the foods we consume, and we continue unaware of its existence. More obviously, humans cause harm or destruction either accidentally, such as in a vehicle accident, or directly with criminal intent and hostile activity. Nature plays an important role as well, with random acts of destruction from violent weather, earthquakes, floods, and volcanic activity. Other natural causes come from space with meteors, asteroids and comets. All in all, multiple hazards often present themselves randomly. And each hazard includes various probabilities of occurrence and levels of severity. These probabilities and consequences present threats to our personal worlds and the organizations in which we belong. While it remains impossible to avoid all hazards all of the time, a well-devised game plan helps us to step through the hazardous minefield of risks with greater chances of success.

Human Intervention

In this truly active and random universe, shotgun blasts often happen. Objects streak out into the universe, hoping that random probabilities will bring minerals, elements, water, and so forth to the farthest reaches. Unlike these random celestial activities, humans alone demonstrate a unique characteristic– we interact with, and can alter, our world. We can create opportunities to intercede; and we can change the probabilities to ensure we are *safe enough*.

Norman (1990) writes that we are *unique* in *two* ways from all other life. We perceive and interact with our environment, and we record and therefore pass along knowledge, unlike any other known species. The difference from a universal random activity is the human ability to perceive, think, and communicate artifacts in our world (Norman, 1990). In practical ways, we see this in the way a hunter may use a shotgun to shoot at a flying bird, hoping at least one of hundreds of small pellets will strike the target. With careful technical design and training, the hunter aims a designed grouping of pellets in a studied and practiced direction that improves the probability of a hit. In similar ways we direct our activities to improve the odds of probability whether to achieve success or to avoid harm in a risky world. Several ensuing chapters propose ways to apply our 'two' uniquely human abilities to manage our risks and make our lives safer.

Human Minds Change Probabilities

Considering that the word 'hazard' comes from the Arabic word *al zahr,* or dice (David, 1962), let's consider probabilities of rolling dice. Figure 4.1 lists various 'natural' probabilities for rolling a pair of dice (Bernstein, 1996).

Sum	Probability
2	1/36
3	2/36
4	3/36
5	4/36
6	5/36
7	6/36
8	5/36
9	4/36
10	3/36
11	2/36
12	1/36

Figure 4.1 The probability of each sum when rolling a pair of dice

Throwing a pair of six-sided dice produces thirty-six possible outcomes, from two ones (referred to as snake eyes in the game of craps) to two sixes (referred to as boxcars) (Bernstein, 1996). The possible number combinations between the two dice result in the probabilities found in Figure 4.1. Anyone who spends time at the 'craps' table should pay attention to the above table. Note that the possibility of rolling a seven (key in craps) is six times more likely than snake eyes or boxcars. The reason is there are six possible outcomes from the dice that add up to seven, only one possible outcome for two ones or two sixes. Humans can play the numerical odds, which are natural events. However, if a human wishes to intervene in natural events and change the probability, a weight can be placed somewhere on the cube to enhance the probability of a desired outcome, or lessen it for an unwanted result. While I am not advocating cheating, I am demonstrating the 'possibility' of human intervention in the risk game. The same can be true for safety, where a person can manipulate probabilities by mechanical design; such as installing anti-lock brakes to allow drivers better control in a rapid stopping situation.

Events in our own lives and organizations in which we participate can work along the lines of universal principles of random events if we do not intercede. As in the universe, risk is all around us with various probabilities of occurrence and levels of severity. This presents a challenge to leaders when addressing the question, 'How safe is safe enough?' Our re-birth of awareness since the attacks on 11 September 2001 and the Space Shuttle disaster on 1 February 2003 allows us to design and train ourselves to reduce probability and severity. These efforts add important elements of control over our lives. Design, training, and action require effort from individuals as well as leaders. Yet it all starts with courage and a determination to act. These qualities reduce the present, known risk, and also mitigate the unknown problems by decreasing risky behavior in general. A risk management matrix (discussed in Chapter 5) helps increase awareness of ways to reduce risk, and instill a risk management mindset. As we have learned, our unique heritage–our ability to understand, intervene, and record–can make our organizations and the world safer.

What is Probability?

'...without odds and probabilities, the only way to deal with risk is to appeal to the gods and fates.' Bernstein, 1996

Leaders need to know about probabilities to help *safe-up* their organizations. The probable outcome of an activity, compared with associated internal and external inputs, will ultimately determine near-term and strategic goals. *Probability* is a quantitative, or numerical measure of the chances that an event will occur. The probability of an outcome is the ratio of favorable outcomes to the total opportunity set (Bernstein, 1996). While probabilities find expression in measurements of 0-100 percent, the actual numerical measure is from 0 to 1. Zero means no likelihood of the event, where a probability of 1 means absolute likelihood the event will happen. A .5 probability means a 50-50 chance of the event, such as in the flip of a coin. At some point frequent activity of an event must be analyzed to determine the probability. Probability requires numbers to calculate. Bernstein (1996) states, 'Without numbers, there are no odds and no probabilities; without odds and probabilities, the only way to deal with risk is to appeal to the gods and fates. Without numbers, risk is wholly a matter of gut.'

Frequency of like events provides a historical base from which to work. History, from a purely objective, impersonal perspective, critically demonstrates accurate probabilities. As an example, lets look at a hypothetical group of employees. Their job entails stacking sandbags on pallets. A historical review of the past twelve months shows us the recent probability of a back injury while on the job.

$$\text{Probability of a back injury} = \frac{\text{\# of back injuries}}{\text{\# of employees}}$$

If four employees had lost workdays due to back injuries on the job, the numerator is identified. The denominator is the number of employees, but can also be days worked in the past 12 months, or number of sandbags lifted.

$$\text{Probability of a back injury} = \frac{4 \text{ injuries}}{20 \text{ employees}} = .2 \text{ probability per employee}$$

or

$$\text{Probability of a back injury} = \frac{4 \text{ injuries}}{200 \text{ workdays}} = .02 \text{ probability per day}$$

or

$$\text{Probability of a back injury} = \frac{4 \text{ injuries}}{20,000 \text{ sandbags}} = .0002 \text{ probability per bag}$$

A manager in a mid to large company will most likely employ statisticians, engineers or mathematicians who work probable outcomes from accurate data. While numerical probability measurements appear desirable, risk reduction does not always require them. Subjective probabilities exist, where a manager can determine through experience and judgment that a particular event includes a high or low level of risk. We know that risk exposure for employees on a scaffold 20 feet high includes potentially more severity than on a platform only two feet high. We do not need calculations to tell us to install a guardrail on the high scaffold.

Industry uses similar determinants for nearly every activity. Informative historical data is also available from government sources. A good place to start is the Occupational Safety and Health Agency (OSHA) on the Web at http://www.osha.gov.

Maximizing Minimizing Probabilities

In our journey toward a risk-free, or risk-reduced, world we must look for ways to lower probability and reduce severity as often as possible. In any operation, we can play the odds by choosing the lowest probabilities in all of our activities. This does not always require a complicated process, as logic will guide us in most cases. Human interaction can eliminate hazards, control them, or establish procedures to drive down probabilities of hazardous events.

Let's look again at our example of sandbag stackers. A manager of a small organization may not have numerical probabilities available, yet can determine logically that four back injuries are too many. Examples of implementing risk management decisions include teaching employees proper lifting techniques and providing back braces. Such management intervention in the probable outcome reduces the injury event to perhaps one per year.

The minimized probabilities follow:

$$\text{Probability of a back injury} = \frac{1 \text{ injury}}{20 \text{ employees}} = .05 \text{ probability per employee}$$

or

$$\text{Probability of a back injury} = \frac{1 \text{ injury}}{200 \text{ workdays}} = .005 \text{ probability per day}$$

or

$$\text{Probability of a back injury} = \frac{1 \text{ injury}}{20,000 \text{ sandbags}} = .00005 \text{ probability per bag}$$

In the above example, the manager minimized risk for the lifting event, but to maximize risk reduction of other hazards in the entire sandbag operation, numeric or subjective probabilities must be analyzed for every event. For example, there is a risk of dropping a sandbag on a worker's toe; mitigation would be steel-toed boots. Strapping the stacks to the pallet reduces the probability of a stack of sandbags falling onto a person. To lower risk of a sandbag delivery truck backing over someone, an observer can be posted before a driver selects reverse, and a back-up aural warning installed on the truck. Minimizing risk probabilities at every opportunity maximizes hazard reduction.

Maximizing minimizing probabilities (MMP) should occur in all activities both on the job and off the job, and considers every hazard. An everyday example of common sense risk reduction involves lowering the chances of an automobile accident. MMP happens with headlights 'on', speed reduced, tires with good tread, anti-lock brakes, an inspection process before long trips, and a checklist of these items available for drivers. Each one of these prevention tools lowers risk in a particular area; all items together lower risk even further–that's MMP–reducing cumulative probabilities for each hazard. Realize if one risk remains high, the entire system is exposed to that individual risk level, and MMP is not achieved. We do not need numerical probabilities in this example;

subjective considerations will do the job. We know logically that headlights will help improve the odds that other drivers will see our vehicle, slower speeds are generally safer, good tires give better traction to maneuver, and anti-lock brakes help us to maintain control during rapid deceleration, inspections and checklists find the unknown hazards. Lessening severity also reduces the overall risk assessment. Airbags, seatbelts and appropriate child restraints offer such reductions in severity.

We find risks, both on and off the job, in every activity, not just on the roads. Maximizing our minimizing efforts represents essential responsibility. Allowing one probability to remain unnecessarily high exposes our whole operation to risk. Leaders must ask, 'Are we safe enough if we allow one risk to remain high?' To answer the question many business factors are considered: Are resources available to lessen the risk? Would we have to change the mission to eliminate the hazards? Is technology available to reduce or eliminate the risk? Can we stand the result if the probabilities play out against us? Of course, there are many considerations for managers and leaders to consider when determining *how safe is safe enough.*

Universal probabilities of random events 'will' play out throughout the vastness of space. This happens here on earth as well. Thankfully, we as humans have the ability to alter those probabilities, and to increase our chances of survival. We can do this individually or organizationally; both help to achieve success. It takes awareness of hazards in our world, and an appreciation for the possibilities to reduce risk to ensure we are *safe enough.* At the end of the day, leaders must be aware that it also takes a human choice to attack random probabilities, change them, and ultimately *win the risk game.*

Chapter 5

Risk Management

'Human intervention in natural events can change the odds in our favor to protect life, guard assets, and preserve the environment.' Alston (2003)

We must take risks–it is our nature–and great strides and benefits can occur when we succeed. Any action or activity includes associated risks. While we cannot totally eliminate them, we can *manage* them with attention, monitoring, and interventions. Thus the challenge in pursuing our organizational or personal goals is the requirement to manage the risks involved, and to determine, 'How safe is safe enough?'

The private sector formalized risk management decades ago. In fact, many large corporations incorporated risk management into their cultures to the point that their safety offices are now called 'The Office of Risk Management'. In the U.S. Department of Defense, the Army took up the charge in the mid-1980s to formalize Operational Risk Management (ORM) with a five-step process. The U.S. Air Force picked up on the idea in the early 1990s, slightly modifying the Army's model into a six-step process. Any way you slice it, risk management works, and is here to stay in both the public and private sectors–an excellent tool for leaders in fighting risk.

The average risk management model offers several steps that apply to nearly every possible activity from an individual act to a large corporate operation. In basic form, commonly accepted steps are as follows (AFI 90-901, 2000):

1. *Identify the hazards associated with your activity.*
   (The awareness step)
2. *Assess the risks.*
   (Compare probability with severity)
3. *Analyze risk control measures.*
   (Consider the options)
4. *Make control decisions.*
   (Pick the best option, at the appropriate level of authority)
5. *Implement risk controls.*
   (Act to eliminate or minimize risk)
6. *Supervise and review.*
   (Make sure steps 1-5 still apply and are working)

These are formal steps that take effort. Leaders, supervisors and individuals need familiarity with the process–a whole-team effort. Safety professionals also need training in the specifics to shepherd the process. Understanding each step will enhance the overall effort.

1. Identify the Hazards

Hazards are traditionally defined as any real or potential condition that can hinder performance or may cause injury, illness, death, or damage to equipment or property (AFI 90-901, 2000). Organizations need a system to identify hazards, the first step in prevention methodology. There are many ways to identify hazards:

- Encourage individual *reporting* of hazards with reward incentives. This requires training and encouragement, and publicizing rewards. A hazard report form, shown below, is a good tool.

HAZARD REPORT			
Name:	Phone:		
Work Station:	Date/Time:		
How did you become aware of the hazard?			
Describe Hazard and Location:			
Recommendation on how to eliminate the hazard:			
Have you notified your supervisor?	Yes	No	
Did you warn others to make them aware?	Yes	No	
Did you call the risk management office? 555-1234, ext. 4500	Yes	No	
Do you believe this hazard also exists elsewhere? Yes		No	

- Formal *inspections* can discover unknown hazards (Chapter 9).
- Accident *investigations* can find hazards that caused a loss (Chapter 9).
- *System safety* processes that examine preliminary hazard lists and industry trends with experts who can brainstorm for other organizational threats (Chapter 6).
- *Government* notices and circulars.
- *Trade* magazines.

When a hazard is identified, the next step is to assess the threat.

2. Assess the Risks

Risk assessment generally considers two necessary tenants–probability and severity –presented by the hazardous condition (AFI 90-901, 2000). *Probability* is a main consideration of risk, and the second is *severity* (consequence); an evaluation of both determines if risk is acceptable or unacceptable. Typically, Probability × Severity = Risk (Bahr, 1997). Importantly, both quantitative and qualitative factors determine the level of risk associated with a specific hazard.

Quantitative considerations determine 'possible' asset loss should an accident occur. Uncovering all of the costs is not easy, with both tangible and intangible costs (discussed in Chapter 3) that ultimately affect the bottom line. Lost physical assets, lost opportunity due to wasted time, retraining challenges, and diminished corporate image ultimately result in lost production. However, some determinants in risk assessment do not necessarily relate to money.

An example of a *qualitative* factor in risk assessment is a typical parking lot fender bender. When we are struck by another vehicle it burdens our lives. When it affects our quality of life we assess the risk based on qualitative factors that present an inconvenience to us:

- Wasting valuable time to fill out paperwork
- More time wasted to get estimates
- Paying $500 deductible (although this is quantifiable, it affects quality of life because we cannot buy toys with the money)
- Burden to take the vehicle to a repair shop
- Forever living with the new rattle in the repaired fender.

There are some quantitative factors in the above example, payable primarily by the offending driver or those covered by insurance. We attempt to avoid vehicle accidents in parking lots, however, for qualitative reasons–the consequences present personal burdens. After we have assessed the risk, whether quantitative or qualitative, as 'unwanted' or high, we then look at ways to eliminate or mitigate the risk.

3. Analyze Risk Control Measures

Obviously, once we identify hazards and assess the risks, we must decide what to do about them. Eliminating the hazard is the best option from a system safety viewpoint. If it cannot be eliminated, then we may choose to control the risk by using guards, or reduce it by using procedures, or in some other way by changing the conditions. Leaders must know each risk-control option carries its own cost:

- *Impact on mission*–sometimes a high risk cannot be appropriately controlled at an acceptable cost, causing a change in mission. There may be a time when a control measure 'must' be implemented and at great financial cost, resulting in a diminished profits and altered goals.
- *Increased production or operational costs*–Engineering a safety device may cost tens of thousands of dollars (production); controlling a risk by simply waiting out a hurricane costs 'lost opportunity' (operational).
- *Increased burden*–time and inconvenience to circumvent a hazard (could have financial costs or lost opportunity, but may certainly bear qualitative costs).

System safety engineers, who look at all alternatives from design to procedures to determine the best cost effective option, sometimes determine control options (Chapter 6). However, small organizations have tools available to help with this step in the process.

Figure 5.1 helps portray options to lessen risk. The risk management matrix offers a useful tool to help consider choices to lower probability and lessen severity.

			HAZARD PROBABILITY				
			Frequent	Probable	Occasional	Remote	Improbable
			A	B	C	D	E
S E V E R I T Y	1	Catastrophic	High	High	High	Medium	Low
	2	Critical	High	High	Medium	Medium	Low
	3	Marginal	Medium	Medium	Low	Low	Low
	4	Negligible	Low	Low	Low	Low	Low

Figure 5.1 Risk management matrix

The matrix is a tool that can analyze control measures by placing our activities into known levels of risk. While many versions of risk matrices exist, this commonly used, simple version shows relationships between probability and severity. It also illustrates how human intervention affects risk as we consider our options to ensure we are *safe enough*.

To use such a matrix, we need to first become aware of a hazard, and then determine the level of risk by comparing probability with severity. We should be very hesitant to engage in activity where the risk is assessed at A-1, since it is nearly a certainty that the event will occur and the result will be catastrophic. Essentially, to evaluate control options consider that risk should only be accepted when the *benefit of the activity outweighs the possible negative outcome* should the odds play out in the worst way. This is true for all activities, such as driving, playing, investing, eating, relationships, yard work, professional decisions, and any other personal behaviors. While the matrix often considers data-driven probabilities computed by engineers, realize that we do not always have analytical percentages of probability for our activities. At times subjective judgment must determine the final action.

The matrix in Figure 5.1 is a good tool for assessing risks that we cannot eliminate. It is a good way to analyze control measures by re-evaluating risk after applying a measure. We can reduce risk by changing the conditions of our activity to a lower risk box. If we choose to use such a matrix for personal risk management, we can make it easy by replacing calculations of absolute data with subjective determinations for probability, and determine severity by using gut feelings and life experiences, though they will not be as precise as data based calculations.

Risk Matrix Example

Take the example of driving 70 mph on the freeway. The potential *severity* is catastrophic due to the real possibility of death, but the *probability* is low if driving on a dry road, with good tires and no traffic. Under those conditions the risk assessment is '1E', or low risk. However, risk is dynamic, as conditions change so does the risk level because it moves to a different box. If the speed remains at 70 mph, but traffic becomes heavy, the risk moves to box 1D, or medium risk because of the increased threat of heavy traffic. Add rain, and it moves to 1C, or higher with ice and snow. That is high risk.

To manage *personal risk* in this example, we must evaluate our control options and change the conditions. We can slow down to 50 mph (follow our best gut feelings) and lower our risk to a subjective 2C, for medium risk, or eliminate the risk by pulling over to the side of the road and wait for good weather and less traffic (Note: data-driven calculations are possible using government accident figures for similar conditions). Essentially, if we must travel, we lower our risk by changing to more favorable conditions. This exercise can be complicated with data and calculators during a system safety process, but it does not have to be that detailed to get similar results. Gut feelings and common sense considerations can achieve most of the risk management results we desire. Once the various options are considered, a decision is made.

4. Make Control Decisions

After analyzing the control options, a decision is required that best helps the organization. The decision takes into account several factors:

- *Resources*–safety ideas compete with corporate interests and limited resources.
- *Return on Investment* (ROI)–CEOs always consider ROI. Resources are limited; payback is critical to an organization, prompting the question, 'How much safety are we buying for the capitol we spend?'
- *Impact on mission*–A safety measure can alter strategic vectors. An organization may, for example, close down a chemical division that requires extensive financial investment into safety modifications and invest its resources in 'paper production' in an overseas location if ROI is attractive.
- *Litigation*–A risk control measure that appears favorably in the courts is highly desirable should a court case occur.

After the decision is made, make it happen.

5. Implement Risk Controls

This is the action step. On the surface, this step seems obvious. However, it often takes a dedicated effort to write the checks and make it happen. To ensure its accomplishment, someone must be accountable for the process of implementing risk controls (Vice President for Safety, or Safety Manager):

- Build a game plan with a timeline.
- Make sure all players know their responsibilities to implement the control measure.
- Track the progress of implementation to completion, and report it up the chain.

The implementation step takes energy and commitment. But the process does not end with implementation; the control measure must be monitored.

6. Supervise and Review

Risk is dynamic, always changing. Think of risk as an equation, balanced in different directions. When a condition in the equation is changed, the level of risk changes as well. We must monitor all known hazards, even the ones we control:

- Is the control working?
- Are there new conditions?
- Did the control measure introduce new hazards?

In addition to monitoring, leaders must reassess risks to determine the current risk level, and change control efforts as appropriate.

The six steps of the risk management process work if the organizational structure and culture allow. Identifying, reporting, assessing and controlling risk formally are indicative of a safety-minded culture. However, to win the risk game, and ensure they are *safe enough*, organizational leaders must incorporate risk management operationally, backed by policies and procedures, and adopt useful operational tools.

Operational Risk Management

Well-written policies and procedures blend 'real time' risk management into an organization's operations. An organization can formalize the risk management process to fit its operations. There are several steps to a formal program:

- Form organizational policy that directs use of risk management.
- Develop risk management decision tools.
- Provide initial risk management training.
  - o Emphasize individual responsibility.
  - o Train all members in hazard reporting avenues.
  - o Make all aware of their accountability in risk acceptance.
- Provide refresher safety and risk management training.
- Provide supervisor training.
- After members are well trained in risk management, are aware of how to use available tools, and know their responsibilities–hold them accountable for disregard.

Policy and training are covered in Chapter 9, 'The Safety Program'. Risk management decision tools, however, require more detailed discussion. The risk management matrix discussed earlier is such a tool that helps the risk decision process. More formally, numerical values are added for each hazard type, and decision levels are mandated. Figure 5.2 demonstrates.

This value-based approach helps put risk decisions into perspective quantifiably, and can vary numbers to fit the operation and type of hazard. Probability and severity are often quantified by system safety engineers, but can also be subjectively determined with expert opinion. Normally, safety people will analyze data to determine (quantifiably or subjectively) the appropriate box. In Figure 5.2, one is not restricted to the listed values. If a manager feels the probability is not occasional, but more than remote, intermediate numbers may be used, such as 4.7, 5.0 or 5.3.

			HAZARD PROBABILITY				
			Frequent	Probable	Occasional	Remote	Improbable
			10	8	6	4	1
S E	10	Catastrophic	100	80	60	40	10
V E	8	Critical	80	64	48	32	8
R I	4	Marginal	40	32	24	14	4
T Y	1	Negligible	10	8	6	6	1

Figure 5.2 Value-based risk management matrix

Risk Level	Level of Decision
60-100 (High)	Top Management
30-59 (Medium)	Director of Operations
10-29 (Low)	Line Supervisor
1-9 (Negligible)	Individual

Example (hypothetical)

A-Z Trucking Company studies data over ten years regarding the number of mishaps while driving through a mountain pass. They can analytically determine probabilities and severities, depending on conditions. They might determine over the 10-year period, that statistically each year the following happens:

- December-March (winter weather):
  - o 1 truck has a serious accident per month
  - o 2.2 have moderate accidents per month
  - o 7.1 have minor accidents per month

- April-May:
  - o .002 serious accidents per month
  - o .08 moderate accidents per month
  - o 2 minor accidents per month
- June-September (heavy summer traffic):
  - o .03 serious accidents per month
  - o 2.1 moderate accidents per month
  - o 4.3 minor accidents per month
- October-November:
  - o .0007 serious accidents per month
  - o .02 moderate accident per month
  - o 1.7 minor accidents per month.

Having analyzed the probabilities, leaders in A-Z Trucking can determine that the winter months pose a much greater risk. With the information they have, they can determine in which box to place the risk (combination of quantifiable and subjective determinants). In this example they may determine the following while using the matrix in Figure 5.2:

- In January with 400 truck sorties crossing the mountain pass, one truck will have a serious accident: .0025 percent chance that a serious accident will occur. The subjective question is whether or not it is a remote possibility, occasional or probable. Their data suggests they 'will' lose a truck, so it is probable, and it is catastrophic for that truck involved–the risk value is 80, and the top manager makes the call to launch or not.
- In October, using the same method, the chance of a catastrophic accident is .00000175, or improbable and a line supervisor can launch the sortie.

Consider human intervention. In January, A-Z top management can instill hazard control measures to control the risks as follows:

- Heavy snow–do not launch (the cost of this control measure is lost revenue).
- Light snow–launch, but procedurally restrict the truck speed to 45 mph on flat roads and 25 mph on curved roads, and use chains (the slower speeds equates to fewer sorties–lost revenue).
- Clear weather–launch unrestricted, be vigilant for black ice–take tire chains just in case.

Such control measures alter the probabilities to less than .0025 in January, but the cost of the measures may reduce revenue. The payback is not losing more revenue due to an accident. While matrices are an excellent guide to help managers make risk decisions, there are other operational risk decision tools that have merit. One such tool is an *operational phase risk assessment worksheet*.

Risk Assessment Worksheet

This instrument helps in specific operational activities to assess the level of risk, and choose the appropriate level of supervision and 'accountable' person to accept the risk. Let's revisit A-Z Trucking Company to see how the worksheet works.

A-Z has many trucks, hundreds of drivers, and a variety of routes–some easy, some difficult. Prior to departing the loading dock, a driver and/or supervisor complete the worksheet in Figure 5.3.

Pre-departure Risk Assessment Worksheet							
Item	Low Risk	Pt	Medium Risk	Pt	High Risk	Pt
Driver Experience	> 1 Year	0	6 months to 1 Year	10	Less than 6 months	30
Route Difficulty	< 200 mi, Flat	0	201-500 mi., flat	5	> 500 miles, flat	15
	Mountainous	10	Mountainous	20	Mountainous	30
Driver Rest	Well Rested	0	Rested	5	Fatigued	25
Enroute Weather	Clear	0	Rain/wind	10	Severe/snow/ice	30
Departure Time	0700-1500	0	1500-2200	15	2200-0500	25
Truck Condition	New	0	2-3 years old	5	6-10 years old	15
			4-5 years old	10	> 10 years old	25
Cargo Weight	< 50 % capacity	0	51-90% capacity	10	91-100% capacity	20
Driver Personal Problems	None	0	In Divorce	15	Family death,	30
	$ Problems	5	Child in Trouble	15	Layoffs Ahead	25
Driver Familiarity with Route	> 5 Times	0	2-5 Times	5	First Time	10
Driver Record	0 Violations	0	1- 3 Violations	5	> 3 Violations	10
Season	Spring/Fall	0	Summer (Traffic)	10	Winter	15
Totals						
Grand Total		+		+	=	
< 90	Low Risk	Driver accepts mission, shows results to Supervisor, info only				
91-159	Medium Risk	Driver shows Supervisor, Supervisor makes launch decision				
160-239	High Risk	Supervisor elevates launch decision to Director of Operations				
> 239	Extreme Risk	Director of Operations elevates decision to Top Manager				

Figure 5.3 Risk assessment worksheet

The driver simply circles the appropriate point values and adds up the totals. If the risk is low, the driver makes the decision to accept the mission, but shows the results to the line supervisor to keep his boss informed. As the assessed risk increases, a higher level of management must make the risk acceptance call. The level of decision can be detailed in organizational policy and guidelines.

The risk assessment worksheet offers an excellent gauge of risk, but a side benefit is that it culturalizes safety. Individual risk awareness is elevated as people go through the process. If it is company policy to fill out such a worksheet, it shows the deeply held belief in safety that members will embrace. Morning checklists that increase individual awareness can also help leaders achieve this result, and helps answer the question, 'Today, are we safe enough?'

Morning checklists are cheap and easy management tools to achieve a safety mindset. They should not be long and boring, but short, simple and to the point. They can also assign risk points (subjectively) as a risk indicator.

Morning Personal Risk Assessment Checklist:

- Am I well rested?
  - Yes   0 pts
  - No   10 pts
  - Sort of   3 pts
- Am I feeling well?
  - Yes   0 pts
  - No   10 pts
  - Sort of   3 pts
- Am I trained for today's task?
  - Yes   0 pts
  - No   10 pts
  - Sort of   3 pts
- Am I distracted by a personal problem?
  - Yes   10 pts
  - No   0 pts
  - Sort of   4 pts
- Do I have the proper personal protective equipment for my task?
  - Yes   0 pts
  - No   20 pts
  - Sort of   10 pts
- Do I sense conflict with management or fellow employees?
  - Yes   5 pts
  - No   0 pts
  - Sort of   2 pts

Assessment

- 0-20 points = Low Risk   Have a good day!
- 21-40 points = Medium Risk   Be careful, inform supervisor
- 41-65 points = High Risk   Inform supervisor, change conditions

Such checklists are easy to build, and achieve awareness at the start of the day. They help leaders form a culture where people think through aspects of their human condition and personal contribution to the mission. The result is efficiency above and beyond the safety objective–everybody wins.

Risk management does not have to be difficult or complex. Sometimes data-driven analysis is required, and top-level discussions are needed to manage organizational risk. Often, risk management is simply looking for hazards and then removing them. In any case, organizations have a responsibility to follow a risk management process to protect its people, its assets, and the environment, and to ensure they are *safe enough*.

The six-step process discussed earlier in this chapter is a good way to accomplish the task of risk management. Risk management matrices are another way to help consider risk mitigation options by changing conditions, and altering risk. Operational risk assessment worksheets are excellent for a last minute, real-time determination of risk, and to make the acceptance decision at the appropriate level. Lastly, morning safety checklists can heighten risk awareness of individuals, and instill a safety mindset for the organization. These tools are available to all leaders, and work as a profit multiplier to improve the bottom line. Organizations, large or small, must take steps toward survival–risk management achieves that end.

Chapter 6

System Safety: Designing Out Risk

'Executive-level managers will determine safety boundaries in any system safety process.' Alston (2003)

When I teach a system safety course, I tell my students that the first step in system safety is to define the system. In doing so, a central question will drive the process, 'How safe is safe enough' (Bahr, 1997). Top management determines the answer when it has a strategic impact. Given free reign engineers and safety professionals can do wonders to safe-up a system. However, senior leaders must account for fiscal constraints, company image, strategic goals, and various other concerns–in a nutshell, leaders weigh the risks to the organization. Risks in varying degrees are inherent to all activity, and all systems require activity. When answering the question of *how safe is safe enough*, leaders at all levels need to realize that risks must be addressed and eliminated or mitigated to an acceptable level for mission success and organizational survival.

While risk management gets much attention, and rightfully so, it is akin to the more comprehensive process of system safety. Many principles are the same, but system safety requires more focused energy to find the unknown unknowns in single and multiple failures, and considers human interaction with machine. Human interaction not only refers to operators, but all levels from top management to line maintenance, accentuating the comprehensive characteristic of system safety.

System safety assures a system is as safe as possible or 'acceptable' for people, equipment, and the environment. Risk management is similar, but focuses on tradeoffs to eliminate or lower risk in targeted operations. In either case, it is assumed risks must be taken, and therefore the acceptance of the possible outcome should risk mitigation efforts fail. As always, the benefit of the activity must outweigh the risks. This understood, the system safety concept is simple (Bahr, 1997):

- Identify hazards (knowns and unknowns).
- Eliminate or control the hazards.
- Mitigate residual risks (engineering or managerial controls).

While the concept is simple, the process is sometimes arduous. Prior to such an undertaking, system safety engineers must first 'define' what hazards are before they attempt to identify them. Primarily, they are focused on *hazards of concern*–those that threaten the system.

Defining Hazards

A primary tenant of system safety is to define a hazard. A simple definition of a hazard *is a condition that can cause injury or death, damage to or loss of equipment or property, or environmental harm* (Roland and Moriarty, 1990). Engineers and safety professionals must look at the possible failures in normal operations and added complications of multiple failures to determine the real extent of a hazard. A failure is not necessarily a hazard, but becomes a hazard when the end result has enough severity to warrant concern. The risk associated with a hazard is a product of severity and probability of occurrence. More precisely: Risk = Frequency × Magnitude (Bahr, 1997), commonly referred to as Risk = Probability × Severity (or consequence). Often it is severity that renders a risk as unacceptable, but it is compared to the benefit of the activity to ultimately determine acceptability.

Two examples bring clarity to the severity versus benefit decision when determining concern for a particular hazard. The chances of lightning striking a person are low, but the consequence is death if a strike occurs; risk mitigation requires one to take cover in such a storm. The severity of an airline crash is also death and event probability also low, yet millions of people readily accept the risk because the benefits of air travel outweigh the risk. While defining hazards of concern is a theme in system safety, there are team characteristics and critical key steps that put system safety in a world of its own.

The tasks are complex and require great effort from an experienced team. A system safety team normally includes system safety engineers who can determine event probabilities and associate severities. Other common members of the team include those with historical knowledge on similar systems, operators from within the system, experts in the field of study, and certain stakeholders, but normally centers on system safety engineers. In some cases the process can be more subjective than quantitative, but typically involves data searches and statistical probabilities in an attempt to intervene in an accident's chain of events. The system safety process can be very complex depending on the system. For the purposes of this book, to help *executive-level risk management*, there are several 'basic' key steps in the process to give leaders and managers an appreciation for system safety:

1. Define the system.
2. Identify hazards.
3. Assess the risks.
4. Eliminate, control, or reduce risks.

1. Define the System

Initial efforts should focus on defining the system by determining the boundaries. The word system does not imply solely a mechanical system, but includes human interaction at multiple levels, support structure, corporate goals, operating environment, and many other contributing factors.

The System

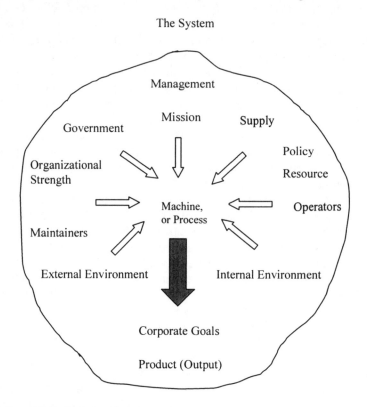

Figure 6.1 Defining a System

A major part of defining a system is to determine what management wants; what is the CEO's vision? By pulsing responsible leaders, engineers can better understand the strategic vector and organizational goals to determine how safe is safe enough. *Safety is only as important as management allows it to be*, and is often shaped by available resources. Once the safety boundaries are drawn, the system safety team can take into account acceptable risk before approaching hazard reduction. The key step in the process, however, is hazard identification.

2. Identify Hazards

System Safety teams often conduct exercises to study accident scenarios of concern to identify initiating, or triggering, events (Bahr, 1997). Triggering events do not always start the accident chain of events, but bring them into play. An accident chain of events may actually begin with executive decisions to cut costs or accept risks. Those are links in the chain, and are factors when a triggering event initiates the chain of events. A fatigue crack can initiate a failure in a structure or part, but may be the result of other factors such as poor maintenance procedures that inadequately inspected for fatigue or corrosion. A 'crack' may also be the result of a deficient design that allowed inadequate metals to form the failed part. Identifying initiating events will find the possibility of a fatigue crack, and work backwards to ensure better metals are used, inspection procedures are in place, and that management does not compromise safety. During the exercise several steps take place to find hazards of concern.

- A *preliminary hazard list (PHL)* is reviewed to consider historical data of typical hazards of a system, sub-system, or associated system. Historical data offers lessons learned of triggering events that can be designed out or mitigated during the current system safety study.
- Along with the known possibilities, some unknowns can be uncovered through *brainstorming* of experts in the subject matter. Great minds of a particular field and fields of associated concern join together to think through the possible outcomes in every part of the system.
- Next will come *list building*, where the team lists hazards of concern, the ones that must be controlled or avoided, such as those leading to loss of life, valuable assets, or revenue.

Hazards are many with varying degrees of risk, and time is limited, requiring the team to focus on hazards of concern. Using the example of an aircraft hydraulic system to determine concerns, a system safety team can identify the following:

| *Little Concern* | *Concern* |
| Hydraulic gauge | Hydraulic failure |

A failed hydraulic gauge is easily replaced and causes no system damage. However, hydraulic failure can cause major damage, huge costs, disruption in operations, and injury or death. During the system safety exercise, there must be a determination of triggering events that could cause the hazard of concern, in the above case–hydraulic failure. At the same time, prevention measures are explored to prevent the hydraulic failure triggers, or in some way mitigate the hazard. Some trigger events are considered in Figure 6.2.

Trigger	*Mitigation*
Over-pressure	Relief valve
Tube chafing	Object clearance
Tube failure	Use stainless steel or titanium tubing
Connectors/fittings/torque	Procedures, torque wrenches, supervision
Arcing	Design of insulation/electrical components
Maintenance damage	Procedures, inspections for dent stress

Figure 6.2 Triggering events that lead to hydraulic failure

Initiating events are many, but most are known and listed in preliminary hazard lists. The system safety team aggressively searches for unknown hazards and adds them to the list of concerns, along with the possible triggering events. Once hazards are known, a critical step in the process is to assess the associated risk.

3. Assess the Risks

Determining Probability and Severity. A central driver while listing triggering events is to determine probabilities, and ascertain consequences, or severity. Quantifying probabilities and severity requires analysis of historical data, as well as WAGs (Wild Associated Guesses) or SWAGs (educated WAGs). This step is difficult because historically engineers do not design things to fail, but cost and available technology may prevent designs that do not fail. Historical data is riddled with various designs that were driven by a blend of upper management decisions marred with varying degrees of risk acceptance that were *not safe enough*. Due to lack of standards, some data do not provide an adequate baseline to determine sound probability. However, historical data can provide some relative constants if boundaries of a data search are well defined. We know, for example, filaments will fail over time, oil will breakdown, moving parts will wear out, load bearing structures can fatigue or be over-stressed, and all things corrode over time. And, of course, people make mistakes that can cause premature failure of a part. Engineers can use the historical data, or test-driven data, to determine failure probabilities. Realizing that a part of the system may fail, consequences can then be determined to help managers make decisions. An 'event tree' is a common method to examine possible consequences.

Event Trees

To help affirm the known hazards, and find as many unknown hazards as possible, event trees help fulfill the system safety process. These tools examine initiating events, projected mitigation efforts, and determine damage states with respect to consequences of the failure. A part of the procedure is to determine what might happen should a mitigation effort fail, and what consequence follows. Importantly, a determination of consequence helps ascertain level of concern. It is essential to know if the consequences are catastrophic, critical or negligible, and the total dollar value (Bahr, 1997).

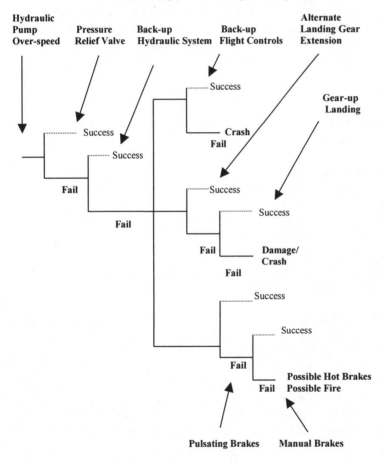

Figure 6.3 Event tree

Let's revisit the aircraft hydraulic system. Looking at the various hydraulic components and back-up safety measures in Figure 6.3, this event tree follows the path to the right as conditions warrant. The worst outcome is a possible crash if the triggering events are not mitigated. If a safety device fails, the engineer follows the 'fail' line to the lower right. If successful, the dashed line on the upper side safely ends that particular activity–the hazard is blocked and the incident is over. However, as each component fails, the event tree follows the failure, or consequence path to the next step. Ultimately, if the governor and relief valve cannot resolve the over-pressure situation, the hydraulic system fails. If the back-up hydraulic system also fails, loss of aircraft control leads to a crash. If flight controls still work, the system still requires an alternate gear extension and use of manual brakes, which complicate landing and have possible significant consequences. If manual brakes fail the next step is aircraft loss of control and runway departure procedures will be followed. All in all, the event tree allows system safety engineers to examine all of the possible outcomes. In the process, they also examine possible failure modes.

Failure Modes

Failure modes are ways a part or component can fail. Bahr (1997) lists typical failure modes that should be considered as:

- Premature operation
- Failure to operate at the prescribed time
- Intermittent operation
- Failure to cease operation at a prescribed time
- Loss of output or failure during operation
- Degraded output or operational capability
- Other unique failure conditions as applicable, based on system characteristics and operational requirements or constraints.

Essentially, all failure modes are considered in the system safety process. Greatest concern is given to those that are possible *single-point failures*. A single-point failure is one that shuts down an entire system. A spark plug on a lawn mower is such a part, where if it fails the motor quits. The consequences for the mower and its operator are nil. Consequences are sometimes catastrophic when a single-point failure in chemical production plant results in an explosion. While single-point failures are the predominate concern, the system safety process must attend to all hazards of concern–and in the end, eliminate or control them–otherwise it is not *safe enough*.

4. Hazard Reduction

There are many parts to system safety; this chapter offers merely a broad-brush view of the process for leaders. Approaches to the process can vary depending on the boundaries of the system, but the end result is the same: hazard reduction. While elimination is preferred, hazard reduction is the primary purpose of the system safety team, and involves three levels:

1. Design out the hazard.
2. Control mechanisms.
3. Procedural controls.

1. Design out the hazard. This is the preferred option when possible because it eliminates risk of a particular event completely. The methods can range from quite simple to very complex. An example of a simple design is found on most new cars on the market today where ignition is inhibited until the transmission is shifted to park. An easy engineering fix prevents the vehicle from lunging when the key is turned to 'on'. In some cases designs that eliminate certain risks are undesirable due to cost, performance limitations or new risks that may arise from a new design. Consider the risk of blown tires at high speeds due to low tire pressure. An engineer can design a solid rubber tire that eliminates the risk of a blow-out, but presents negatives in initial cost, operating fuel costs due to weight of the new tire, new engineering requirements for axle and suspension design to accommodate the weight, and new threats of rubber chunks breaking free. In this example, a control mechanism is preferable, discussed below.

2. Control mechanisms. Safety devices are common risk reduction tools when hazard elimination is not feasible or practical. Relief valves, guarded switches, warning devices (aural or visual), and redundant components to back-up a primary system are common types of control mechanisms. Following the above example of tire blowouts, one control mechanism to help avoid the risk is a pressure sensor. Since the greatest hazard is found in under-inflated tires, the sensor alerts the operator that the pressure in a particular tire is low, enabling the driver to correct the problem before an accident occurs. An element of risk still persists because human interface is required to install, maintain, and respond to such devices. Procedural controls may still be needed to encourage operators to check pressures to avoid the risk.

3. Procedural controls. Management may opt to reduce a hazard by directing adherence to safety procedures, rules or policies.

Examples of Procedural Controls

- Use of checklists
- Supervisor review of work
- Training (Supervisors and individuals)
- Rest periods
- Equipment time changes
- Wear of protective equipment
- Accountability for failure to comply.

There are a host of other rules and policies available. *Procedural controls* are often selected for budgetary reasons, and are less costly up front. Expediency is another reason to use procedural controls; management can write a rule or guideline for immediate use by employees. Though often used, this is the least preferred hazard reduction option because people make mistakes and do not always follow rules. Other human factors can impede performance, such as fatigue, task saturation, judgment, miss-prioritization of tasks, perceptions and the like. Even highly trained professionals are subject to the fallible human condition (Reason, 1990) that sometimes leads to procedural failure, resulting in an accident. Removing the dependency for human interaction by designing out the hazard is often the preferred option for optimum hazard control if mission, resources and technology allow.

Hazard reduction is the name of the game in system safety. It is a difficult task since many parts contribute to a failure. Understanding hazards, and events that put them into play, are principal concerns for the system safety engineer. However, safety engineers should not limit their search for hazards only to component failures. The system can fail from inadequately written procedures and poorly run safety oversights. Take, for example, the Space Shuttle *Challenger* accident in 1986. The Rogers Commission that investigated the accident found that the disaster was caused by a failure in the aft field joint between two lower segments of the right solid rocket motor (Report, 1986). On the surface, it sounds as though it was a single-point failure, and to a mechanical extent, it was. However, other failures led to accident. In the *Challenger* accident, as well as other high-visibility accidents, it is apparent that latent errors pose the greatest threat to the safety of a complex system; root causes are often present within the system long before active errors are committed (Reason, 1990). It was discovered with *Challenger* that the solid rocket booster joint was poorly designed, and it was a well-known deficiency. The Report also pointed out that the 'system' led engineers to attempt to fix problems in-house rather than informing 'accountable' decision makers. The communication system between engineers and management was poor, and information was often incomplete or misleading. Ultimately, political pressures to launch, poor design, inadequate communication, and bad weather (icing) all contributed to one of America's greatest disasters. Were NASA's safety processes *safe enough* at the time? Adhering to solid safety policies and sound system safety discipline can prevent such calamities, and have a positive impact on the bottom line. As well, it is important to consider the added system safety benefits to an organization that transcend safe operations.

Benefits Beyond Safety

A sound system safety process can prevent failures in system design, but must also examine the human procedural inputs, information flows, and various interfaces of decision makers to make sure the system is *safe enough*. Leaders must familiarize themselves with the system safety process to realize the full slate of benefits to the organization. The results of the team efforts go beyond safety and find operational efficiencies for the corporate good. While pursuing a safe system, the team often uncovers inefficiencies that need repair. Some 'gained' efficiencies beyond safety are:

- Improvements in communication flow
- Organizational structure
- Assessments in organizational health
- Time management improvements
- Process improvements
- Economies of more efficient systems
- Training enhancements
- Corporate image.

Organizations need the system safety process. It can benefit any sized organization, using either a large or small team, and can help answer the persistent question, '*How safe is safe enough?*' Society as a whole profits from the process, where our daily lives are 'safed-up'. Recreational facilities, efficient vehicles, electronic devices, lines of communication all have safety and efficiencies derived from the system safety process. It is a 'winning' process that leaders and managers need to be aware of, and use.

Chapter 7

Organizational Risk

'Risk is inherent to the jobs we do in the Air Force. Our safety challenge is to identify hazards, mitigate risks, and make smart decisions on what risks we will and won't accept. This applies to everyone and every activity from driving to work, to executing a combat training mission at Red Flag, to a 30 hour bombing sortie.' General Ralph E. Eberhart, as Commander, Air Combat Command, Safety Message to the Command, September 1999

General Eberhart made the above statement as part of a command-wide Safety Message for the Air Combat Command. In the message, he directed a Safety Down Day, a day of no flying in which all ranks would take time to reflect on how they do business and focus on safety issues. The General knew how important it is for the CEO (in this case a 4-star general and commander) to influence the course of safety. His effort was successful; an adverse trend in human factors-caused mishaps turned around in outstanding fashion, beginning a two-year success story of driving down accident rates to record lows.

Leaders at all levels, not only CEOs, must realize risks attack organizations from many directions. Risk describes a measurement of uncertainty. Uncertainty exists about whether a loss will occur, when it will occur, what will be the size or severity, how frequently will it happen and how this loss will impact on the business enterprise (Unknown source). Only after we grasp these uncertainties, and attack the hazards that threaten our mission, can we begin managing them as an organization, and adequately address the question, '*Are we safe enough?*'

Hazards with associated risks confront every organization. Known hazards we attempt to eliminate or control; others remain unknown. First, we need to know what hazards confront the organization. Chapter 9 outlines safety program essentials, one of which is a reporting system to identify previously unknown hazards. The importance of reporting hazards is obvious: we cannot eliminate a hazard if we do not know it exists, nor can we answer the question, '*Are we safe enough?*' Another essential for hazard identification is a safety inspection system that uncovers physical hazards as well as training or procedural deficiencies. An analysis program also helps by examining data to discover adverse trends and problem areas (Chapter 9). Additionally, as reviewed in Chapter 6 on System Safety, a brainstorming session with a preliminary hazard list helps discover the spectrum of threats to the organization.

We find organizational threats, or hazards, in many forms. These warrant continual scrutiny. Typical hazard categories found in all industries are (Bahr, 1997):

- Collision
- Contamination
- Corrosion
- Electrical
- Explosive
- Fire
- Human factors
- Physiological factors
- Loss of capability
- Ionizing radiation
- Temperature extremes
- Mechanical
- Pressure.

While the above list reveals typical hazards to an industrial organization, one hazard inherent to essentially all organizations irrespective of their purpose or mission remains–the 'well-intentioned' human hazard. Well-intentioned humans can make mistakes or succumb to individual inefficiencies. Fallible people can write deficient training standards or failed policy directives; they can fail to give proper training, guidance, or supervision. Leaders must address the human factor when looking at organizational risks–the human element determines *how safe* we are.

The importance of the *human factor* in the cause of mishaps was introduced in the foreword to this book. Essentially, humans often stand in the way of achieving 'zero' accidents. When we consider that all accidents are avoidable, it suggests that a zero accident rate is possible. So then, why do we still have accidents? We, the human factor, are the primary reason. James Reason (1990) asserts:

- Fallibility is part of the human condition
- We cannot change the human condition
- We 'can' change the conditions under which people work.

We are fallible because we have certain limitations and physiological frailties. We can suffer from fatigue, complacency, poor judgment, spatial disorientation, optical illusions, panic, distractions, depression and many other symptoms that make us susceptible to cause accidents. However, we make attempts to change work conditions to assist our frail selves. For example, a truck driver has anti-lock brakes to assist in safe stopping to compensate for poor judgment or slow reactions. Also, rules are written to make sure truck drivers are well rested and drive a reasonable number of hours per day to avoid fatigue (we rely on the drivers to follow the rules). All in all, we cannot change our human fallibility, but we can change the conditions around us to help make us *safe enough*.

Organizations are made up of 'fallible' individuals. One person can adversely effect an organization by accepting unacceptable risk. As stated earlier, we coexist with risk individually and corporately, requiring our careful concern. An important element of risk management is to accept risks at the appropriate level of management to ensure only risks that are acceptable to the corporate body are allowed. Organizationally, there are three major levels of risk acceptance:

- Strategic level
- Operational level
- Individual level.

Strategic Level

Strategic decisions are made by senior leaders and affect the corporate mission, types of products, methods of production, and promote safety policy, structure, culture, and climate. Strategic decisions on risk must consider the frailties of the individual, and not rely on hope that probabilities will not play out; individuals ultimately make the organization a success or failure.

> 'Our folks need to know that they have the ability to call a "knock it off" if the risks are getting too high.' General John Jumper, Chief of Staff of the Air Force, Safety Message to Commanders, June 2002

General Jumper delivered a direct yet simple strategic decision regarding individuals. He realized that his 'can-do' Air Force has the courage to push their limits to the maximum; they seldom say 'can't'. While his statement clearly demonstrated the view from the top, the strategic level, it was directed at the operational level where those implementing policy sit. They needed to hear this message first-hand from the 'Chief' as it funneled downward within the organization to the Air Force workforce. Strategically, Type A individuals, as found in the U.S. Air Force, need to know they can back away from unnecessary risks without being viewed as weak. It is strategic because the CEO communicated organizational policy that gave individuals risk management power, and encouraged mid-level management buy-in and support, and allows the operation to be *safe enough*. Importantly, all organizational levels must support risk management to make it work.

Operational Level

Operationally, an organization breaks down risk acceptance decisions into various leadership levels:

- *Director of Operations*–Determines risk acceptance that can risk a life, bring the operation to a halt, or cause major environmental or equipment damage. Launching a truck fleet during adverse weather is such a decision–someone could die, and corporate image is at stake.
- *Mid-manager*–Decides on, and enforces wear of proper protective equipment and HAZMAT safety procedures. Decides whether one truck should drive 'one more' line with borderline tread on a tire.
- *Line supervisor*–Accepts risks that are less likely to occur, but could cause a localized problem. Should not accept the risk of launching a truck with a bad tire while carrying hazardous materials; may accept risk of a new driver driving a difficult route for the first time if he or she felt certain the driver could safely handle it.
- *Individual*–Make risk decisions that affect their daily operation, but do not put their well-being at risk or risk functionality of the production line.

If any level of risk acceptors notices a bigger hazard, they elevate the decision to the next higher level of management, who will determine if the activity is *safe enough*. If an individual sees that the tread on a tire is badly worn, he tells the line supervisor, who (if not cleared to make the decision) then tells the mid-manager who makes the decision. The decision level is based on accountability. If a truck does not drive, the company loses profit; if it drives and there is an accident, there is also lost profit: there is accountability in either case. The appropriate level must decide on risking the mission with a bad tire or losing revenue by not driving. Importantly, note that hazards are not solely mechanical, but more often pertain to people. On the human level, as in Gen Jumper's message, the individual who identifies himself or herself as personally 'not good to go' must be dealt with without retribution, addressing any needs, and with help to return them to their former, peak level of performance.

Individual Level

The *individual level* of risk acceptance remains incredibly important. Operational activities are comprised of numerous 'individual' decisions daily. Organizational leaders need to consider the role of an individual very seriously. One person can inappropriately accept a risk that burns down a plant, or detonates combustible material, or shuts down the operation.

This means individuals need detailed training on their responsibilities regarding risk, and safety in general. They should know how to protect themselves and not risk their well-being. They must know how to make risk decisions that avoid placing production in jeopardy. Realizing we cannot practice for every situation, decision-making training pays off, when unexpected risks arise. People need to know the reasoning process of risk management to make it work, because they will ultimately face an unforeseen dilemma. In the end, individuals need to know *how safe is safe enough*.

Example of decision-making

I have seen the following scenario. An individual was tasked to drive to a stranded company truck to change a flat tire in order to get products delivered 'on time' to the customer. Once there, the individual noticed there are no 'chocks' available for the tires, and the truck was situated on a hill. The hazard is gravity and truck mass, which may cause the truck to roll off the tire jack. The severity of the risk: the individual could be crushed. The person was faced with a risk management decision. He could hope the truck was stable and would not roll off the jack, or he could return to the shop and get the tire chocks to ensure his safety. He wanted to be a 'can-do' person and get the truck to the destination on time, but assessed the risk as high due to the high probability and severe consequences.

Dilemma. If the above individual returned to the shop to retrieve the chocks (placing individual safety over organizational production) and then changed the tire, he risked customer dissatisfaction due to a late delivery and potential future business. However, if he attempted the task without chocks, and the truck rolled over him, he risked injury or death. Organizationally, and perhaps unknown to the individual, there were other risks. One was the possibility of lost productivity and truck revenue for several days during an ensuing accident investigation. Another organizational risk is directly financial–the monetary payouts for the accident would decrease corporate assets. Also, negative press would harm company image and affect future business. Luckily, this individual recalled intense training on company policy to not accept this kind of risk, and the multiple layers of reasons why, so he returned to get the chocks. The moral of the story is that *individuals make up organizations*, and *individuals make risk acceptance decisions every day*; they are each key players on the team.

'I encourage you to solicit inputs from all ranks, across the entire spectrum of your unit activities, to identify risks and ways to eliminate or control those risks. Once these are identified, unit personnel should develop a plan or procedures to keep these potential accidents or incidents from happening, and as a final step educate unit personnel on the results of this effort.' General Ralph E. Eberhart, as Commander, Air Combat Command, Message to commanders for a safety focus day.

In his directive above, General Eberhart reminds his field commanders that the whole team is important. Anyone can make a safety input. The subtle message in the General's statement conveys that organizations are attacked by threats from many directions. To survive the assault, the entire corporate body must be involved and work as a team. Leaders are important, but so is each individual. Ultimately, individuals make daily risk decisions within an organization–the CEO, director of operations, line supervisor, machine operator, secretary and the janitor.

If individuals make up an organization, and at the end of the day make the risk acceptance decisions, then leaders must understand that some people are more likely to accept risks than others. Paradoxically, risk takers are important for organizational success. Recall that *all activity bears risk*; if risk is unduly avoided then success is reduced. Organizations must march toward their mission–operations require activity and thus risk. In their quest to answer the question, '*How safe is safe enough*', organizational leaders should understand the human propensity to take risks.

Why do people take risks? Perceptions that may impact risk acceptance vary from one person to another. Slovic et al. (1979) have determined there are certain factors that affect how people perceive risk (Bahr, 1997).

- *Is the risk voluntary or involuntary?* People more readily accept risks they choose; skydiving is an example of a personal choice to accept significant risk. However, if a risk is forced upon them, people often oppose the risk, even if the risk is low. If a prison is being constructed in their neighborhood, members of the community may fight in the courts because of the unwanted and 'unasked for' risk in their backyard, even though the risk is low.
- *Are the consequences catastrophic?* Perceived catastrophic consequences raise the concern for risk. Since large numbers of people die in an aircraft crash, it is sometimes perceived as high risk, even though the chance of dying in an aircraft crash is 1 in 7 million (Barnett, 2002).

- *Are the consequences dreaded or common?* Nuclear contamination from a nuclear facility is dreaded and raises risk perceptions to a level where people will fight to remove the hazard, though the risk is minimal. Whereas risk from a chemical plant is higher with the larger number of chemical accidents, but since it is a more common occurrence people more readily accept the risk.
- *Are the consequences certain death or uncertain?* A company driver may risk driving on a flat icy road with little or no traffic, where death is possible but less certain. However, he or she would not risk a mountainous icy road with a deep cliff off of one edge of the road, where the consequence is certain death.
- *Are the consequences immediate or delayed?* An aspect that affects risk perception is the chronological nearness of the risk. Consider smoking cigarettes. Though the risk is high for getting cancer, you will not get lung cancer tomorrow, it will not be for many years; the risk is easier to accept.
- *Are the risks technologically controllable?* The controllability factor pertains to whether an individual feels they can personally control the risk. Many people would not bungee jump because they cannot control the events—once they jump, they are in the hands of the gods.
- *Is the risk new or old?* Traditional risks are more readily accepted. Driving your car is a common tradition in the United States, but presents a high risk due to severity of the consequences, yet we often do not think twice about it. A new technology risk that has not yet become a tradition, such as space travel, may be safer overall, but it is new and seems untrustworthy; a perception that may skew risk assessment.

Essentially, we all have varying degrees of risk tolerance and risk acceptance. Organizations need risk takers, but they need smart risk takers who do so at the appropriate time and level, and who avoid unnecessary risks. Studies suggest this ability to make *pragmatic* decisions is a distinguishing characteristic of successful fighter pilots in the U.S. Air Force (Dillinger, 1994).

While some accepted risks come from human choices, risks also come from other sources. Organizations are faced with risks across the spectrum of activity, and must consider all avenues of risk. Environmental risks, and organizational risks that include structure, behavior, culture and climate can all present risks to an organization (Ginley, 2002).

Environmental Risks

Environmental risks greatly affect the organization's bottom line. The environment of an organization is most easily defined as anything outside the boundaries of the organization (Nelson and Quick, 2000). However, both internal and external factors can impede production with little fault on the part of the organization. I say little fault because good risk awareness can resolve most risks, so the organization is sometimes accountable for 'not seeing' the threat. Figure 7.1 reveals some of the internal factors and external *environmental* factors that pose risk to an organization's assets and its very survival.

Figure 7.1 Environmental risk factors

Environmental Factors

Environmental factors cause pressure on an organization. How do these factors become safety risks? Pressure at some point will lead to distraction, worry, concern and unrest–human factors that cause accidents. In this light, leaders at all levels must monitor such external environmental factors to determine overall organizational climate. If the environment is static, there is little uncertainty causing stress on the organization. However, a dynamic environment causes great uncertainty, and may influence the actual design of the organizational structure, where an organic structure will allow for adaptation (Nelson and Quick, 2000).

Internal Factors

Internal factors equally present risks that must be managed. These include one's work ethic. Work ethic reflects responsibility and care for corporate assets. Poor work ethics present risk to the organization. Sub-optimum ethics not only affect production, but also infect others' attitudes with lack of concern. Sound work ethics toward safety are important within an organization–they help seek out risks and eventually eliminate them. Added to work ethic, organizational philosophies and policies help construct a healthy culture toward risk management, or a sick culture if philosophies and policies are weak and misguided. Such internal factors help contribute to the shape of the organization's overall structure.

Organizational Structure

The *organizational structure* can harbor hidden risks based on poor policies and weakly written guidance. Some organizations are quite complex with numerous breakouts by countries, regions, divisions, functions and business units. Communications become intangible hazards when people in complex organizations do not receive hazard information or preventative guidance. *Organizational processes* must allow effective communication and a well-organized safety training program for new personnel, annual refresher training, and safety training for supervisors (Chapter 9). Regarding safety and risk management, organizational leaders must clearly define roles, responsibilities and accountabilities that leave no doubts for any member. Formal rules in the form of written policies, regulations, procedures and guidelines shape the organization's structure (discussed in Chapter 9). A clear safety structure will positively affect safety behavior. To best manage risks, safety must be embedded in the key *dimensions* (Nelson and Quick, 2000) of the organizational structure:

- *Formalization:* Official rules, regulations and procedures.
- *Centralization:* Degree to which safety decisions are made at the top of the organization.
- *Specialization:* Safety jobs should be narrowly defined and require unique expertise.
- *Standardization:* The degree to which safety is accomplished in a routine fashion.
- *Complexity:* How safety interfaces with different types of activities within the organization.
- *Hierarchy of Authority:* Vertical safety chain across levels of management.

Sound organizational structure is vital to the risk game. Clearly defined responsibilities and lines of communication are key; involvement from the corporate top is essential, and leaders at all levels play a role. Structure, inevitably, affects the overall complexion of organizational behavior.

Organizational Behavior

Organizational behavior defines corporate stability, but may present risk. Nelson and Quick (2000) define organizational behavior as *the study of individual behavior and group dynamics in organizational settings*. Essentially, organizations are comprised of individuals who must work together to achieve the mission. Factors that internally impact an organization's behavior are:

- Personal needs of members
- Personal goals
- Goal congruence for the organization and its members
- Organizational incentives
- Cooperation
- Conflict.

A myriad of other associate factors also apply, but risks to an organization depends on the health of the above list. If individual needs are not met, risks surface from distraction, depression, complacency and lack of dedication. Divergent goals lead to disunity. When organizations lack incentives, people lose focus and loyalty. If cooperation is lacking, resentment may cause internal pressures that pose risk. Conflicts present distrust, disloyalty, unhappiness that distracts people from their assigned tasks. Organizational behavior is a product of the organization's general health, and a by-product of its *culture*.

Organizational Culture

Organizational culture is a pattern of basic assumptions that are considered valid and that are taught to new members as the way to perceive, think, and feel in the organization (Nelson and Quick, 2000). Cultural strengths and restraints help to keep the culture stable. *Cultural strengths* are habits of thought and behavior and structural characteristics that have served and will continue to serve the organization if not threatened by change. *Cultural restraints* are deeply held assumptions that condition and restrain thinking about the future, people are content with the present. The overall cultural depends on the levels of cooperation, individual development, and group commitment to organizational goals. *Organizational risks* that affect safe operations can rise from poor function, ambiguous policy, disjointed processes, undefined roles, unclear responsibilities and inconsistent accountabilities. These risks affect the overall *safety culture* of the organization.

Interest in the term 'safety culture' is traced back to the Chernobyl nuclear accident in 1986. Since then, numerous definitions of safety culture have been put to pen. Among the many, several commonalities are determined (Wiegmann et al. 2002). They are:

- Safety culture is a concept defined at the group level or higher, which refers to the shared values among all the group or organization's members.
- Safety culture is concerned with formal safety issues in an organization, and closely related to, but not restricted to, the management and supervisory systems.
- Safety culture emphasizes the contribution from everyone at every level of an organization.
- Safety culture impacts members' behavior at work.
- Safety culture is usually reflected in the contingency between reward systems and safety performance.
- Safety culture is reflected in an organization's willingness to develop and learn from errors, incidents, and accidents.
- Safety culture is relatively enduring, stable, and resistant to change.

Considering these commonalities, Wiegmann et al. (2002) propose a global definition of safety culture:

Safety culture is the enduring value and priority placed on worker and public safety by everyone in every group at every level of an organization. It refers to the extent to which individuals and groups will commit to personal responsibility for safety, act to preserve, enhance and communicate safety concerns, strive to actively learn, adapt and modify (both individual and organizational) behavior based on lessons learned from mistakes, and be rewarded in a manner consistent with these values.

Further, there are five global indicators that help measure safety culture within an organization:

- Organizational commitment
- Management involvement
- Employee empowerment
- Rewards systems
- Reporting systems.

The level of depth that these components are engrained into an organization reveals the overall culture of safety–leadership contributes to the depth. Organizational well-being depends on strong, healthy organizational and safety cultures. While these cultures are enduring, an organization is also affected by the more temporal impacts of the organizational safety *climate*.

Organizational Climate

Organizational climate differs from culture in its lack of permanence and stability. The safety climate in particular, is a 'snapshot' of employees' perceptions of a current environment or prevailing conditions, which impact upon safety (Mearns et al., 2000). Based on common themes in various definitions of safety climate...Wiegmann et al. (2002) derived the following definition of safety climate:

> Safety climate is the temporal state measure of safety culture, subject to commonalities among individual perceptions of the organization. It is therefore situationally based, refers to the perceived state of safety at a particular place at a particular time, is relatively unstable, and subject to change depending on features of the current environment or prevailing conditions.

Safety professionals and organizational leaders cannot overlook the importance of safety climate within their organization. The safety climate can fluctuate with real-time influences, such as new leadership, organizational changes, and environmental impacts. Safety managers, and corporate mangers in general, must be vigilant for climate changes, realizing the constant fluctuation. Through diligence and effort, the climate can possibly be controlled with incentives, training, and participative processes.

At the beginning of this chapter, the quote from General Eberhart pointed out that *risk* is inherent to an organization; the safety challenge is to identify hazards, eliminate or mitigate risks, and make smart decisions on what risks we will and will not accept. This applies to everyone and every activity, and determines if we are *safe enough*. Risk management requires prevention methodology: identify hazards, report them to appropriate people, assess the risks, choose the best solution, and monitor. Sometimes we must accept risk, but must accept it at the right level–strategically, operationally or individually. In management's effort to win the risk game, sound organizational structure and solid corporate processes that achieve prevention methodology will, in the end, effectively manage organizational risks, and provide a comforting answer to the question, '*Are we safe enough?*'

Chapter 8

Personal Risk Management

'Human error is a very large subject...' Reason (1990)

To err is human; to intellectually rise above other species, seek out hazards, and systematically eliminate or control risks is also human. Successful leaders know how to manage risk in their lives, but also must understand the importance of Personal Risk Management (PRM) among the organization's members. Obvious needs exist for 'on-the-job' risk management, and we manage those risks in several ways; personal protective equipment is one, corporate procedures are another. Also, at work most people are supervised by someone who has a stake in their well-being–productivity. Supervisors need to know, however, that their stake in an employee does not end when the quitting whistle sounds.

If an individual is injured or killed off the job, on a weekend, the organization still loses productivity and a line supervisor must train an 'unknown', new employee. The organization is not *safe enough* if its members are dying while off the job. Leaders must be mindful that hazards confront organizational members 24 hours per day, 7 days per week whether at work or on personal time.

Individuals with different talents and skills make up organizations, and those talents and skills come into play when confronting risks. Sometimes people work alone, or may not always be under the trained eye of a supervisor. On the job and off the job, members comprise the most important element within an organization–any loss ultimately affects the mission, and brings into question whether the organization is *safe enough*. The challenge to corporate leaders remains how to encourage individuals on the job, or alone on the weekend, to protect themselves from the many threats. In training programs, managers help individuals form habit patterns that fight risk and educate them to manage the risks in their lives in several key ways. Individuals need to:

- Know their personal responsibility to avoid unnecessary risk.
- Identify risky behavior, and control it.
- Use a written Personal Risk Management Guide.
- Utilize a simplified version of risk management.
- Embrace personal risk reduction rules.

Personal Responsibility to Avoid Risk

Managers and leaders should make sure all members in an organization know their importance to the mission. If someone belongs to an organization, they bear responsibility for its success, and they must be told that in no uncertain terms. In the military, for example, the responsibility runs deep. First, military members who are killed on the weekend let themselves down by cutting their lives short. Secondly, they emotionally harm their friends and loved ones, who miss them. Next, they failed their military unit; others must take on the deceased individual's duties along with their own, and accomplish those actions required in response to the death. Essentially, their death diminished their branch of service, which must find and train a replacement at great cost of time and money. Lastly, they let their country down in a time of war against terror–all members of a military organization are important and needed. Whatever the organization, we all have responsibilities to stay alive, and stay healthy–leaders must teach safety responsibility to all organizational members.

We have heard the saying, 'It's my life, and I can do what I want. It doesn't involve you.' When accepting risk, we should always remember that it is not 'just' our own lives involved. What we do affects others. Just as our investment practices, whether good or bad, affect an entire family, either positively or negatively; accepting unnecessary risk in other activities does the same. If someone is irresponsible in risk assessment, and becomes paralyzed, many people are burdened emotionally and financially for many years. Ultimately, lack of awareness or irresponsible risk acceptance kills, harms, or inconveniences family, friends, co-workers, acquaintances, and those in our paths. Irresponsible risk acceptance is a betrayal of trust. Ultimately, we have a responsibility to protect ourselves, and begin by assessing our own risky behaviors.

Identify Risky Behaviors

Our complex personal universes share many interconnecting worlds. We each initially present a foundation, and branch out to the reaches of our influence. How we behave in our private lives often affects how we behave in our professional settings, associations and social interactions. Our newfound quest for security since the terror attacks on the World Trade Center, the Pentagon, and Flight 93 in Pennsylvania, requires an increased awareness of our entire environment, but must begin within ourselves. Rarely an easy task, it is quite difficult to evaluate one's self honestly for risky behavior. Self-examination causes stress, and is often a battle between the *ego-ideal* and *self-image* (Nelson and Quick, 2000). Ego-ideal is the embodiment of a person's idealized (wished for) perfect self, with no distasteful qualities, where the self-image is how one truly sees himself or herself. The comparison often causes internal conflict called 'ego-dystonic' or, 'I'm not what I thought–as good as I thought– and it hurts'. However if a person wants to look within, it is possible to identify the prominent risky behaviors without much stress, such as propensities for:

- Speeding on a freeway, weaving in and out of traffic
- Drinking and driving
- Use of power tools without eye protection
- Working on electronic devices without locking out electrical sources
- Hiking, swimming, or boating alone.

There are, of course, many possible examples of risky behavior. In this exercise, it is important to note that the tendency to take risks or avoid them is only part of behavior toward risk. Risk taking is influenced not only by an individual's tendency, but also by organizational factors (Nelson and Quick, 2000). Pressure to perform, deadlines, and a task oriented organizational culture rather than a safety-oriented culture will cause members to push beyond their normal propensity to avoid risks.

Individual propensity to accept risk is difficult to determine. It changes with other variables such as age, job, and marital status. Some key determinants for risk taking are discussed in Chapter 7. However, since we know external factors can modify an individual's behavior, we can potentially increase or decrease certain behavior if we better ourselves. For example, let's look at some revealing elements of our daily reality. These three questions provide a behavioral glimpse into our selves, to see if we are personally *safe enough*:

1. Where do I put my money? People value money, and how they spend money offers a window into their behavior. Revelations of risky behavior are found in one's receipts: alcohol purchases, late night bar tabs, turbo-chargers at the auto parts store, motorcycles (crotch-rockets), and so forth are indicators of accepted risks, with sometimes poor consideration of the consequences. When discovered, list risky receipts on a list of behaviors to change or eliminate.

2. What do I do when no one is watching? People reveal who they are by what they do when alone, when no one is there to hold them accountable. Self-examination of behaviors when alone offers insight into a person, but requires self-honesty. Some risky behaviors when no one is watching are: drinking and driving, not buckling a seat belt, failure to wear goggles with power tools, and stealing money from the snack fund at work. These can be many, but are eliminated from one's life with awareness, desire and effort. Add these to the list of personal behaviors to change.

3. What do I do when I'm with my peers? Peers can have an unhealthy effect on a person. In many cases, young people will accept unnecessary risks with a group, risks they would never consider alone. Much of the time, these situations involve alcohol, which makes risk acceptance even more pervasive. Individuals must seek out personal threats when groups encourage unnecessary risk taking, and have the courage to eliminate or avoid those hazards.

Once risky behaviors are determined, add them to the list. Examine those behaviors individually, and review the benefits for change. Begin each question below with, 'If I change this risky behavior':

- Will I live longer?
- Will I be healthier?
- Will I save money or my career?
- Will I preserve future opportunities?
- Will I show respect for family, friends, organization and myself by changing?

The next step is to decide how to control those risky behaviors.

- Talk to friends about changing behaviors.
- Change risky friendships.
- Personally take charge and make changes for improvement.
- Get professional help for uncontrollable behaviors.

Then, lastly, take *action* to change adverse behavior. Determining risky behavior increases awareness of threats to one's life; awareness is key to winning the risk game and being *safe enough*. While challenging, global awareness of all threats is possible when sincere individuals use available tools.

Personal Risk Management (PRM) Guides

We must learn 'how' to manage the threats in our lives. One tool both individuals and managers use to assist in PRM is a Personal Risk Management Guide. A PRM guide walks people through the risk management process for a specific activity. Flipping through the pages, individuals review typical concerns in their upcoming activity, be it hiking and camping, boating, scuba diving, or any other activity, and make the proper risk decisions. Also, managers can go over the topical information with subordinates to ensure hazards of the activity are thoroughly considered. A PRM Guide walks people through the six steps of traditional risk management. The first step in PRM is always, '*A commitment to win the personal risk game*'. Each of us must want to minimize threats to our lives.

The six steps in review:

Step 1: Identify the hazards
Step 2: Assess the risk
Step 3: Analyze risk control measures
Step 4: Make control decisions
Step 5: Risk control implementation
Step 6: Supervise and review.

An example of the process brings clarity to the value of such a guide. The following is an excerpt from the U.S. Air Force, Air Combat Command's Summer Personal Risk Management Guide (Air Combat Command, 2002).

Safe Boating

Step 1: IDENTIFY THE HAZARDS. Let's look at the hazards associated with safe boating:

- Weather (storms, wind, swells, tides)
- Location (lakes, rivers, oceans)
- Boat condition/drain plug open (cracks/holes in hull, leaks)
- Motor condition (old, broken, unreliable)
- Fueling (vapors, spills, explosion)
- Exceeding people/equipment limit
- Slippery/wet surfaces
- Lost (no Global Positioning System (GPS)/compass/map)
- Stranded (sandbar, reef, rocks, submerged trees)
- Speed
- Alcohol
- Safety equipment.

Step 2: ASSESS THE RISK. Assess the impact of each hazard in terms of potential loss and severity:

- Injuries, damage, and lost equipment due to severe weather and water conditions
- Mishaps on lakes, dangerous currents in rivers, and rapid tidal changes for inlets, etc.
- Boat sinking due to taking on water from damaged hull or open drain plug
- Drifting due to motor breaking down/flooding out
- Fires due to fuel vapors or spills

- Capsizing due to exceeding the load limit of people and or equipment
- Falling overboard, hypothermia, or drowning
- Traveling in the wrong direction
- Hypothermia, dehydration, sunburns, or drowning
- Loss of control, collisions, capsizing, or running aground due to excessive speeds
- Intoxication, impaired judgment, unnecessary boat maneuvering
- Overnight cabin safety, running air conditioners (carbon monoxide poisoning).

Step 3: ANALYZE RISK CONTROL MEASURES. Once you have identified the hazards and assessed the associated risk, you should decide on some controls that can be employed to reduce or mitigate the hazards:

- Attend local Coast Guard safety course
- Start with a good safety briefing prior to heading out
- Listen to the National Weather Service for the day's forecast and plan accordingly. Cancel boating trip if inclement weather is expected. Ensure all safety equipment is ready and available
- Familiarize yourself with lakes, rivers, and inlets before attempting to navigate on your own
- Ensure equipment is inspected. Have boat motor and any other equipment serviced routinely
- Ensure personal flotation devices are available for all individuals, flares, and first aid kit
- Use extreme care when fueling. Clean up any spilled fuel. Don't let anyone smoke or have open flames near gas tanks. Try to keep gas tank area well ventilated
- Travel at speeds safe enough for water conditions
- Don't overload the boat with people or equipment
- Ensure all occupants wear properly fitting US Coast Guard approved floatation vest
- Keep an emergency kit onboard that contains food, blankets, sun block, fresh water, and flares.

Step 4: MAKE CONTROL DECISIONS. Accept the risk, avoid the risk, reduce the risk, or spread the risk. Do not make dumb decisions.

Step 5: RISK CONTROL IMPLEMENTATION. Once you select appropriate controls, use them! A plan is only good if it is followed.

Step 6: SUPERVISE AND REVIEW. As always, the situation is subject to change quickly. Monitor the situation and adjust as necessary to keep things under control. Summer is a great time to have fun and we all deserve a break every now and then. From now on, use risk management to make your summer fun, memorable, and safe. No one wants a summer outing to turn into a tragedy!

An outstanding tool, PRM Guides offer a simple review of the hazards, mitigation options, and decision-making steps that help individuals start with the right mindset. Awareness remains the key to safety, reviewing the boating guide will help to heighten awareness in a valuable member of an organization. Other activities in the above-described PRM Guide include:

Sky Diving	Scuba Diving/Snorkeling
Swimming	Roller Blading
Fishing	ATV Operations
Boating	Lawn Care and Gardening
Power Tools	Operating Vehicles
Hiking and Camping	Motorcycling
Mountain Biking	Basketball
Softball	Jet Skiing
Horseback Riding	Soccer
Golfing	Bull Riding
Barbecuing	Home Repairs
Fireworks	Trampoline Safety

PRM guides present a useful tool for individuals, by themselves or along with supervisors, to review hazards and walk through the six steps of risk management. It helps individuals to answer the question, '*How safe is safe enough?*' A simple tool, it represents a hands-on way to manage personal risk. However, people are often confronted with hazards in their daily activities where PRM guides are not available, and supervisors are not present. People often die alone, when they accept unnecessary risk. Thus, time dedicated to heighten awareness, judgment, and improving timeliness and accuracy of personal decision making is always time well spent, and can be found in a simplified daily tool.

A Simplified Version of Risk Management

Several years ago, when I took over as the Director of Safety for the Air Force's Air Combat Command, it dawned on me that most people could not name the six steps of Operational Risk Management discussed in Chapter Five.

Even I had a difficult time recalling exactly what the steps were while out on the flight line; I could not expect the basic fighter pilot or crew chief to rattle them off. After discussions with the risk management expert on my staff and with several other safety professionals, an idea materialized for a tool to simplify the risk management program. This tool was central to the Command's success in reducing mishaps for a record setting year. While never becoming an official Air Force program, the idea provided an excellent informal guide using a three-step process that even an Airman Basic could keep in a hip pocket: *ACT*.

Assess the environment for hazards.
Consider options to eliminate or control the hazard.
Take appropriate action (to minimize risk).

'ACT' can apply to any task to help prevent a mishap, thus keeping hazards at bay. ACT can work for people on the production line, or when enjoying summer activities off the job; people do not need a scientific study comparing probability and severity for every activity in their lives. Individuals already know the basics of being careful, such as wearing a seatbelt to reduce the severity of injuries should an accident occur. Normally, common sense can guide the average person.

As an example, a pilot can look at bad weather at the intended field of landing and mentally apply *ACT*. Assess the landing environment for such hazards as low visibility, strong crosswinds, ice on the runway, or heavy rain. Consider options to either land immediately, hold for 20 minutes until the poor weather passes, or divert to an alternate runway at another field. Lastly, Take action as appropriate to eliminate or control the risks. *ACT* is an easy tool that an individual can use in both personal and work activities.

The ACT model offers an uncomplicated, easy tool to help people manage risks. Typically, we understand that a chance of an accident is higher during a rainy day. We do not need to compute probabilities. We simply *assess* the environment to determine if hazards are present, *consider* options to eliminate or control those hazards, and importantly, *take* action as appropriate to zero out or minimize the risks.

Along with 'ACT', some rules people can carry with them are helpful. As a squadron commander at Holloman Air Force Base in 1996, I often preached key rules that helped our young members manage risk in their lives.

Six Personal Rules that Reduce Risk

Risks surround us, some obvious, some subtle. Achieving awareness of each and every hazard that might befall us presents an ever-constant challenge.

Encouraging people to use six practical rules will help walk them through the daily minefields of risk.

RULE 1: Maintain awareness: awareness is key to risk reduction. Vigilant awareness initiates and sustains the primary step in basic risk management and the system safety process. Eliminating hazards requires first routing them out. Individuals need to develop a mindset to canvas their environments for threats. Reviewing a PRM guide prior to an activity helps to jog one's mind of hazards and avoidance. In reality, however, people will not have access to such a guide, so the three-step 'ACT' process can help. One can start with a few questions:

- In what I'm about to do, what are the threats that cause harm?
- How can I eliminate those threats?
- What is the residual risk, and is that risk necessary?
- Does the benefit of my activity outweigh the leftover risk?
- Am I prepared to live with the consequences if probability plays out against me?

Points four and five are important questions to us all. People minimize exposure to most risk by taking the time to increase awareness, but ultimately, they must make a decision on whether or not to accept residual risk and ask, 'What is the benefit of my activity?' And, of course, 'Can I live with the consequences?'

RULE 2: Realize you must live, and die, with the choices you make. We are products of a lifetime of choices. Choices on college, marriage, investing, careers, and fun have shaped who we are today—we must live with those choices. This theory carries over to risky behavior as well. Some people have a propensity to take unnecessary risks. Are they prepared to live, and possibly die, with the consequences if the probability of the hazard plays out? They most likely ignore the consequences, because they have not assessed their risky personal behavior in the earlier 'Step 2'. Speeding on the highways, drinking and driving, using power tools without protective gear can all result in a person living in a nightmare for the rest of their lives. Would they, however, take such risks if their mother were watching?

RULE 3: Do what you are doing as if Mom were watching. To examine personal behavior, a good question to ask is, 'Would I be doing what I'm doing if Mom were watching?' If the answer is yes, then the behavior is most likely okay. If the answer is no, stop the activity, it is risky behavior. A person would not drive 100 miles per hour on a wet road if Mom were in the front seat. Moms value their children and want their safety, and are a good guide for personal behavior.

RULE 4: Question things that do not seem right. A good rule of thumb is, 'If something doesn't seem right, it probably is not'. Gut feelings are good guides for behavior. They are a compilation of life experiences that help decision-making. By questioning things that do not seem right, hazard identification and risk assessment are taking place. That is 'assessing' an environment for hazards, the first step in 'ACT'. Make it a point to follow and cultivate your instincts, they are usually correct.

RULE 5: Look beyond your emotions when making decisions. This is an important rule. Generally, people take huge risks when guided by emotions. It is critically important to build and practice a mindset that filters through emotions to follow basic steps of risk assessment. Some emotionally driven behaviors, such as uncontrollable anger, require professional counseling. Pursuing such counseling is the 'T' in *ACT*: Take action to minimize risk. A simple example is road rage, where people allow their anger free reign; this increases their personal risks and risks for others. Another example is when we allow a loved one to talk us into a risky situation, one that we know we should avoid. We can ask ourselves two questions:

- If they love me, why do they put me at risk?
- If I love myself, why do I allow it?

It is important to control emotion-based, impulsive decisions, and to construct a habit-pattern to use careful steps to identify hazards, consider options, and select the best way to eliminate or control personal risks.

RULE 6: Gather courage to do the right thing. It is not always easy to do what we know is right. Controlling personal behavior often requires courage. People face pressures from peers, relationships, finances, and time that present risk–these take courage and discipline to overcome. An example of a courageous decision is to respect ourselves enough to say 'no' to friends who place us in harm's way by drinking and driving, or by attending risky drug-using parties.

I have seen other situations where young Air Force members work a 12-hour shift on Friday, then drive all night to reach a weekend destination. Though fatigued, they felt they could not waste the time or money to stop at a motel for rest–the consequences were too often catastrophic. Does the benefit of saving time and money outweigh the risk–loss of life? No. However, people nation-wide accept this type of risk every day. Courage and discipline come into play in this example in two ways: 1) make the financial decision to spend money on a motel, which requires forfeiting another desired activity due to having lower funds; or 2) cancel the road trip due to high risk of falling asleep at the wheel. Rules 1-5 can help root out hazards, but PRM takes courage, as well as discipline, to do what is right.

In our busy lives, a myriad of topics that affect our personal universes preoccupy our minds. To increase awareness of hazards and associated risks, we must rethink our approach to our daily lives, and our view of the future. Lessons learned on 11 September 2001 obviously require vigilance for terrorist activities. However, our awakening is not limited to only people of terror, but should address all of the hazards we have ignored or disregarded in the past. Leaders need to encourage vigilance in all that their organizational members do to pave the way to *safe enough* status; awareness of hazards in our environment is essential. Part of awareness is to be mindful of the full spectrum of hazards.

Some hazards are physical, ranging from typical obvious ones such as busy traffic, while others are not easily observed as in deadly bacteria. Some hazards may affect our mental well-being, where mental abuse, brainwashing, manipulation, confusing 'con' games and so forth have a negative emotional or behavioral effect. Hazards to one's family also exist, where personal decisions impact loved ones either in a positive or negative way. As well, we cannot overlook professional hazards. One's profession occupies a major part of life, and a job is not immune to hazards and risks; awareness is essential for positive career growth. In our new view of the universe, we must consider hazards come in many forms, shapes and sizes and must not be ignored.

Leaders have the above tools to help their people in safety: Personal assessments, risk management guides, *ACT*, and personal rules to reduce risk are helpful to individuals, and therefore organizations. Organizations are made up of individuals; one unnecessary loss of an organizational member has negative corporate consequences. A review of Chapter 3 on costs of an accident will offer leaders fidelity into the financial ripple-affect of losing someone to an accident. Most injuries and deaths occur off the job, but have the same impact on the organization: lost productivity. Leaders need awareness of their members' well-being, and endorse a personal risk management program that will lead to an organization that is *safe enough*.

Chapter 9

The Safety Program

Officer of the Court: 'Show us your safety program.'
CEO: 'I am proud to show you our safety program.'

Leaders must continually ask, *'Are we safe enough?'* Why? Why is it so important to have a strong safety program? Costs. An organization has a moral obligation to protect its people, but the driving force behind a safety program is the cost of not having one. Chapter 3 discusses costs of losing the risk game, which are reduced to one thing: lost productivity. Direct and indirect costs weaken an organization's ability to conduct its mission and deliver its products. Direct, but delayed, costs are compensatory or punitive damages from litigation, which are sometimes devastating to an organization, especially when a negligent corporate board fails to establish a solid safety program. When an accidental death occurs within an organization, the courts will look closely at the safety program to determine if adequate protective measures are in place. While we do not do safety to impress courts, it sure helps if something goes wrong. Essentially, a safety program must address hazards and mitigate risks, and abide by Government regulations and statutes. However, safety is much more complex, requiring great effort and oversight.

A good safety program has five major parts. Below, the parts are simply stated, but the complexity of each is great. Leaders at all levels must be involved in the process, and should:

1. Write (or approve) a safety policy letter or statement from the CEO, or organizational leader.
2. Set clear safety objectives/goals to help achieve the mission.
3. Develop an organizational safety structure in depth, with qualified personnel, and the required eight key components.
4. Devise a prevention methodology.
5. Help safety compete for resources.

All five parts must support the organization's mission. Simply stated, the mission comes first. However the mission will fail if risks are not appropriately addressed. Therefore, safety is integral to operations that work to achieve the mission. As stated earlier, the CEO (or organizational leader) determines the level of safety in an organization. Therefore, it is appropriate to begin the safety program with a clear understanding of the CEO's belief in safety, which is laid out in a policy letter.

Part 1. The Safety Policy Letter

A leader of an organization must declare his or her policy regarding safety and risk management. It is important for the policy to support the mission, but it must also support a level of risk acceptance. All activities have associated risk, so any operations that support the mission will carry risk. The corporate top must determine the acceptable levels of risk, at least in the broad view. Line managers and supervisors will determine the 'on scene' levels of risk acceptance, until the losses from mishaps begin to affect mission success. When the mission is at risk, the senior executives will expend resources to mitigate risks, or write policy to avoid them. All in all, top executives, and all leaders below them, must get involved in the risk game to ensure success, and it begins with a policy letter.

Safety policy letters can be extensive or simple. Simplicity often gains the best comprehension among all corporate players. Here are a few examples of organizational safety policy letters, with follow-on assessments.

Example

From: The Department of the Army, Seattle District, Corps of Engineers, (Department of the Army 2002).

31 July 2000

MEMORANDUM FOR: All Seattle District Employees

SUBJECT: Seattle District Safety and Occupational Health Policy

1. As the leader of the Seattle District it is my responsibility to ensure that our people are given a safe and healthful workplace. Accidents are an unacceptable hindrance to accomplishing our mission for the Nation. I expect all leaders to employ the risk-management process in their daily decisions to ensure that resources are conserved and we remain the flagship District.

2. As dedicated employees in this organization, it is up to each of you to follow safe work practices and to assist your coworkers in doing the same. Safety begins with each team member setting the example. Please recognize your

value to the organization and take the necessary steps to protect yourself by working safely.

3. As the world's premier engineering organization, the Corps of Engineers has set many impressive standards. One of those is an excellent accident frequency rate for both construction work and O & M work. Another is our reputation for designing and building award winning projects. Designing for occupant safety and health is a key part of this successful recipe. To maintain this standard of excellence, we must continue to develop our workforce through training in life safety.

4. I expect each supervisor to incorporate specific safety-related objectives into the TAPES forms of all employees in key safety positions like shop foreman and quality assurance representative. I expect each team member to join me in a partnership to create and maintain a safe work environment for us all.

/signed/

X

Colonel, Corps of Engineers

Commanding

Assessment

This is an excellent safety policy letter. In paragraph 1, Colonel 'X' establishes that he is in charge, and that he expects all leaders to use risk management in their daily decisions. He also states that accidents are not acceptable. Are they? He needs to say they are unacceptable, however, reality indicates that only accidents caused by accepting unnecessary risk or through negligence are unacceptable. When an organization accepts risk, it must also accept the possible result, which is an accident at some point due to human fallibility, making it acceptable in reality. Perhaps the colonel should state the unacceptability of taking unnecessary risks that lead to accidents. However, by noting that accidents are unacceptable, he conveys a mindset of intolerance for mishaps. Tasking leaders to manage risks is a good starting point, but there is much more to this letter that strives for an organization that is truly *safe enough*.

In the second paragraph, the colonel implores the 'team concept' within an organization, and calls on each individual to begin with personal risk management, and then to help one another. Paragraph 3 establishes the image and the mission of the organization, where safety is a main component. The colonel states, 'Designing for occupant safety and health is a key part of this successful recipe. To maintain this standard of excellence, we must continue to develop our workforce through training in life safety.' Life safety is interesting and important, since risk management is essential on and off the job, in every activity in our lives. In the final paragraph, line supervisors are tasked as team members in a partnership to culturalize safety throughout the organization. The colonel covered all of the bases in a simple safety policy letter, which worked well for his organization. However, some organizations are huge and more complex, such as mega-corporations that must communicate and enforce a safety policy to thousands of members.

Example

The Lockheed Martin Aeronautics Company is a large organization, and a great safety study with a well-established safety program. Lockheed managed, however, to simply state its safety policy so every member in the organization can understand it (Lockheed Martin, 2002).

Lockheed Martin Environment, Safety, and Health Policy

People and the environment–no two issues are more important to Lockheed Martin Aeronautics Company in the performance of its business. For this reason, the company takes an aggressive approach to the development of policies designed to protect employees, customers, contractors, communities, visitors, and the environment from any potentially unfavorable effects of company activities.

By their nature, some of our activities can introduce hazards into the workplace or pose risks to the environment if compliance to established policies is not strictly observed. To this end, the following company Environment, Safety and Health (ESH) policies are fundamentally constructed within the concept of continual improvement. They are subject to an ongoing process of review in order to accomplish specific goals. Lockheed Martin Aeronautics Company protects employees, customers, contractors, communities, visitors, and the environment from the hazards of Company activities, products or services. We:

- Prevent pollution, conserve resources, reduce waste, and recover or recycle resources where economically feasible.

- Maintain a safe and healthy workplace to prevent injuries and illnesses.

- Comply with applicable laws and regulations, and satisfy corporate and customer requirements.

- Minimize significant adverse ESH impacts by integrating ESH management practices into business decisions.

- Integrate ESH management practices into design processes to minimize adverse ESH impacts throughout production, use, and disposal of products.

- Integrate ESH management practices into procurement and property renovation, rearrangement, acquisition, consolidation and divestiture.

- Develop ESH performance objectives and targets to ensure continual improvement of the Environment, Safety and Health Management System (ESHMS) and reduce adverse ESH impacts.

- Respond to employee, community, customer, and regulatory agency concerns regarding potential adverse ESH impacts due to LM Aero activities, products or services.

- Establish pro-active partnerships with regulatory agencies, customers, and suppliers to improve ESH performance and compliance cost effectiveness.

- Provide people, specialized skills, technology, training, and budget to maintain an integrated ESHMS.

- Maintain ESH requirement awareness throughout the workforce and execute tasks using safe, healthy, and environmentally sound work practices.

Much is involved because ESH management involves myriad details and often-complex issues. However, with these straightforward goals and the active involvement of employees, regulatory agencies, customers and the local community, Lockheed Martin Aeronautics Company has earned an exemplary ESH record.

ESH is Everyone's Responsibility!

Assessment

I would prefer that the CEO signs the policy letter, but the fact that it is a declared company policy suggests it is embraced by the corporate top. While well structured to help accomplish their mission, it also wards off litigation attack on their safety program. The policy shows aggressive activity and social concern in mishap prevention. In fact, they have incorporated the words, '...the company takes an aggressive approach to the development of policies designed to protect employees, customers, contractors, communities, visitors, and the environment from any potentially unfavorable effects of company activities.' Pretty much everybody is listed. Note they mention 'activities'. Recall that all activities carry risk. This letter acknowledges potential unfavorable effects from company activities, thus suggesting they know they are in a risky business, and that eventually an accident will occur, but they imply all precautions are pursued and taken–that they are *safe enough*.

Lockheed Martin's Policy covers many bases. The letter leads off stating that people and the environment are the two 'most important' issues in the performance of their business. In reality, the company mission is predominant; the senior executives, however, are acknowledging that safety is integral to operations, joined at the hip to accomplish the mission. They drive home the reality that the company has great risks, and established prevention policies must be strictly observed. The letter gives a laundry list of things that 'we', the organization as a whole, will do, such as prevent pollution, maintain a healthy workplace, comply with laws, respond to concerns from employees and the community, and provide skills and training to enhance safety. The 'we' concept is key to a great safety program, promoting a whole-team approach to prevention. Lockheed has the team approach entwined throughout their policy proclamation, ending on subtle but important concept.

Lockheed's safety policy ends with the sentence, 'Environment, Safety and Health is Everyone's Responsibility!' This is a simple sentence with great importance. If a member of the organization is held responsible, they are more likely to use prudence in their behaviors. Along with responsibility must come accountability with punitive measures as outlined in Chapter 2 of this book. Lockheed Martin managed to capture all the key elements for a company safety policy, which implies keen corporate insight and responsibility in the delicate matter of safety. An even larger company has a similar approach as Lockheed, with an interesting addition.

Example

General Electric (GE) is one of the largest companies in the world. Yet their safety policy is simple and straightforward, with additional encouragement for employees to be on the lookout for hazards and safety violations. Looking first at GE's policy statement, note they acknowledge weeding out 'unreasonable' risks in their list of core requirements, confirming the reality that some risks are acceptable.

GE Environmental, Health and Safety Policy Overview

GE is committed to achieving environmental, health and safety (EHS) excellence. This is a responsibility of management and employees in all functions. GE will strive to provide a safe and healthy working environment and to avoid adverse impact and injury to the environment and the communities in which we do business. Our programs must combine clear leadership by management, the participation of all employees and functions, and the use of appropriate technology in developing and distributing GE products and services (General Electric, 2002).

Core Requirements

- Comply with all relevant EHS laws and regulations.
- Create and maintain a safe working environment and prevent workplace injuries.
- Reduce waste, emissions and the use of toxic materials.
- Appropriately assess and manage our EHS risks.
- Eliminate unreasonable risks from our products, activities and services.
- Address site contamination issues in a cost-effective and appropriate manner.
- Respect the environmental rights and interests of our neighbors.

What GE Employees are Urged to Watch For

Unsafe activities and conditions, such as:

- Failure to use personal protective equipment (shoes, safety glasses, hearing protection, etc.)
- Unlabeled chemicals
- Exposed or unsafe wiring
- Blocked fire exits
- Unsafe driving or failure to wear seat belts
- Working in high places without fall protection
- Working beneath heavy, suspended loads, or improperly using cranes
- Working on electrical or powered equipment without following appropriate lock-out, tag-out procedures
- Failure to comply with health, safety or environmental regulations and procedures
- EHS complaints from employees, customers or neighbors
- Deficiencies noted by government inspectors
- Unreported environmental, health or safety hazards or accidents

- Failing to respond promptly to concerns about possible product safety issues
- Missed opportunities for reducing waste and toxic materials.

Assessment

GE captured the essence of a 'whole-team effort' toward safety and risk management. Such safety policy letters or statements are important to an organization. They set the corporate tone when conducting activities, and lead to more efficient mission accomplishment. Leaders must get involved in their safety programs, and begin by establishing the overall safety policy, one that clearly states what it takes to be *safe enough*. Once a policy is established, leaders must turn their attention to clear, achievable safety objectives.

Part 2. Safety Goals and Objectives

As with policy letters, safety objectives can be complex or simple, and simplicity is often the best course. The most important consideration regarding objectives is to ensure they are measurable. Safety is sometimes difficult to measure; a safety manager cannot determine how many lives were saved, or accidents prevented. However, an organization's safety performance can be compared to national averages, or to its own historical track record on safety to determine if they are safe enough. There are some different approaches that organizations can take.

Most safety objectives focus on training, because an organization can control and track the training effort. A goal in training can be aggressive to train 100 percent of employees, and easily measured. Such goals are:

- Train all new employees prior to starting work (and semi-annually) on:
  - o   Company safety policy
  - o   Hazards in the work area
  - o   Use of Personal Protective Equipment (PPE)
  - o   Safety procedures and techniques
  - o   Risk management and decision authority
  - o   Hazards in the community while off the job, such as poisonous snakes, boating hazards, dirt-bike activities, local driving conditions, etc.
- Train all supervisors in safety semi-annually on:
  - o   The entire above list
  - o   More advanced risk management decision-making
  - o   Human Factors that cause accidents
  - o   Their roll as safety monitor, and their accountability.

Safety training for 100 percent of organizational members costs money and slows production. However, recall the driving cost of a safety program is the cost of not having one. When people are taken from their jobs to get safety training, they will conclude the importance of the subject matter, and their value to the organization. If 100 percent of organizational members are not trained in safety, the organization is not *safe enough*. Training goals enhance mission accomplishment, can be aggressive, and are easily measured.

Sometimes goals are established for a reduction in accidents or lost workdays by a certain percentage on a target date. These are tougher.

Example

A goal for an 80 percent reduction of lost workdays over an 18-month period. In order to achieve that objective, a game plan with key measurements must be established, and a pledge from the corporate top to commit corporate resources.The following is a hypothetical game plan to achieve the 80 percent reduction.

Game plan

1. Determine how many lost workdays due to injuries your organization suffers annually. In this example, 790 lost workdays are experienced annually; the 80 percent goal amounts to a reduction of 632 lost workdays per year within the next 18 months.
2. Research why people are missing work due to injury (complete in 60 days). Determine the hazards causing injuries and the physical/mental results of each. Decide which major hazards, if mitigated, would have the biggest impact on achieving the company objective. Some simplified examples for the demonstration are:

Top 4 Hazards	Result
Inadequate lifting training/PPE	Back injuries, 207 lost workdays
Production of fumes/noise	Headaches, 192 lost workdays
Tripping/falling	Broken bones, 147 lost workdays
Bumping heads on equipment	Head injuries, 107 lost workdays

3. Mitigate hazards. To meet the 80 percent reduction goal, the organization must expend resources to eliminate or reduce the threats to the operation.

Hazard Mitigation With 'Tracked' Time Measurements

Inadequate lifting training/PPE

- Train and supervise all employees on proper lifting techniques (within 90 days). Supervise work area.
- Provide back braces designed to prevent back injuries to all lifters (within 90 days).
- Include a policy statement that any employee lifting without proper training or safety equipment will be help accountable, along with the supervisor on scene.

Production of fumes/noise

- Fund and install a ventilation system that eliminates 95 percent of fumes in the work area (within 5 months).
- Fund an install sound suppression devices on machinery where possible (within 5 months).
- Provide maximum hearing protection and training to all at risk employees (within 60 days).
- Hold employees and supervisors accountable for proper use of protective equipment.

Tripping/falling

- Install guard rails in high threat areas (within 5 months).
- Paint warning lines where applicable (within 90 days).
- Establish policies to ensure an organized and safe work area (within 60 days).
- Train all members on awareness, safe work techniques and company procedures designed to prevent falls (within 90 days). Supervise operations.

Bumping heads on equipment

- Provide protective headgear and training (within 45 days).
- Redesign and/or reposition equipment where possible to eliminate the threat to workers (with 90 days).
- Hold workers and supervisors accountable to follow procedures and wear protective headgear.

After the above steps are completed, the safety staff must monitor and assess progress toward achievement of the safety goal of 80 percent reduction in lost workdays. In a sense, the above game plan is not merely a 'safety' game plan, but an organizational strategy–safety is integral to operations, and instrumental to mission success. The plan requires dedicated resources and effort from the corporate top to the line worker. Specific timelines for critical elements of the plan are established. The Chief of Safety must update senior executives on the progress of the plan to ensure the target dates are met and the 80 percent reduction is on track. While the objective will enhance corporate performance in general, another benefit becomes apparent. When senior executives expend resources to protect employees (even if the ultimate goal is mission success), the message filters down through the ranks that safety is important–the CEO has determined the level of safety in the organization, and is willing to pay for it. Such action helps to build the safety culture within the organization, and communicates that leadership is committed to ensuring they are safe enough. However, not all safety goals are as specific as reducing mishaps by a certain percentage; some are effective with broader reaching objectives.

The above illustration lays out specific objectives. However an organization should have overarching broad safety objectives as well. A simplistic yet effective set of objectives is found in the Schlumberger Corporation. Schlumberger has corporate offices in New York, Paris and The Hague, and provides oilfield services, network solutions, and systems integration. Schlumberger's approach leads off with a corporate vision on safety, and then lists simple objectives.

Example (from Schlumberger Technology)

Safety is No Accident

One of the best applications of Schlumberger technology is to make the job safer. A safer job is more efficient and less costly. These benefits are the result of careful analysis of potential hazards followed by the application of technology and technique to eliminate them. This approach leads us to our goal–zero accidents. Doing things right the first time pays off for us, and you, every time (Schlumberger Technology, 2002).

Schlumberger Safety Objectives:

- Accident Prevention Team (APT) approach implemented across the company
- Risk Reporting (RIR) remains the cornerstone of accident and loss prevention
- Driving Monitors installed on all Schlumberger vehicles
- Defensive driver training for all field drivers

- Implement safety campaigns that address key risk profiles of each product line
- Safety training to continue to be an integral part of all personnel development.

   Schlumberger's safety objectives center on safety training and risk identification with implied risk mitigation. Supervision through driving monitors, defensive driver training, and safety campaigns to address key risks are all achievable and measurable. Generally speaking, an Accident Prevention Team (APT) approach attempts to develop a safer work attitude by finding true cause of accidents and incidents, and then communicate that information to all employees. Commonly accepted APT goals are:

- To educate employees in the safest methods of job performance
- To prevent accidents and increase the safety of operations
- To find the cause of accidents and make prevention recommendations
- Provide a peer review to determine human factors in accidents & incidents
- Recognize employee achievements in accident prevention
- To educate employees in correct accident reporting
- To standardize information and make it more precise
- To involve employees in prevention procedures and improve the process
- To protect resources (people, equipment & money).

Assessment

Regardless how an organization approaches a safety program, senior leaders must first establish a safety policy, then set clear safety objectives for the organization at large. While such objectives can be quite complex, they need not be. Often, simplicity works better for total organizational comprehension. If specific reductions in accident occurrences are the objective, them a complex game plan may be in order. However, most organizations require a simple, straightforward set of objectives as listed by Schlumberger. Once policy and objectives are determined, top leaders must turn their focus to an appropriate safety structure that best fits their organization.

Part 3. Safety Structure in Depth

Who's in charge? In a nutshell, the person who is accountable, with authority to commit resources, is the person in charge. I have stated several times, the CEO determines the level of safety, and is ultimately the head of safety. However, there are different levels of safety authority, all help determine *how safe is safe enough.* Mid-level supervisors have line authority, and can make on the spot safety decisions that commit or effect some resources, they are in charge within their span of control. All managers and supervisors play a valuable role in the risk game. Generally, for administration of the safety program, the CEO establishes the level of safety, and a Vice President or Director of Safety administers the program. Safety managers and professionals help run the safety program.

Safety managers typically have little direct authority over operations, because they do not have accountability. Managers who are accountable have authority. Often, a safety manager cannot stop an operation, but will advise the Director of Operations or line supervisor that a hazard is present, and recommend that operations cease until the hazard is eliminated or controlled. The manager with decision authority, and accountability, makes the final decision. However, while safety managers do not normally have direct authority, they do have various conduits of indirect authority (Alston, 2003):

Delegated authority. The CEO, Vice President, or Director of Safety confers their authority to the safety manager. If a safety manager recommends action to a Director of Operations, the Director realizes the CEO or Vice President has delegated safety oversight to the safety professional. Thus, the Director of Operations normally follows the safety manager's advice.

Positional authority. As in the above example, Mid-level managers know that safety personnel have the boss's ear, with safety updates and briefings at corporate-level meetings. Positional authority often translates to power.

Authority derived from laws and regulations. When safety people say how things should be, government statutes back them up. Under this mandate, safety people can direct compliance with State and Federal laws.

Authority derived from reputation and education. Safety managers and representatives are the most knowledgeable people on safety rules and practices. When they speak, others listen. The reason for this is safety managers have experience, and are normally required to have safety education.

There are numerous safety college degrees available, and endorsed by the Board of Certified Safety Professionals (BCSP). Many organizations require their safety personnel to have a Certified Safety Professional (CSP) certificate. Associates degrees in Safety and Health are also available, determined by the

BCSP (2002), requiring at least 12 semester hours in course material such as:

- General safety
- Safety management
- Safety compliance
- Safety technology
- Occupational safety
- Safety relating to a particular industry (such as construction, insurance, manufacturing, transportation, health care, etc.)
- Safety and health communication and/or training.

Safety employees who do not qualify for the CSP, which requires a bachelor's degree in any field or an associate degree in safety and health, can consider pursuing the Occupational Health and Safety Technologist (OHST) certification. Those in construction can get a Construction Health and Safety Technician (CHST) designation. Both are nationally accredited certifications and the qualifications for them are less stringent than those for the CSP. While neither requires a degree, many people use the OHST and CHST as stepping-stones in their career while also pursuing appropriate academic credentials (BCSP, 2002). In the final analysis, safety professionals have earned their derived authority.

The organization must understand the importance of safety decisional authority. But authority is only one piece of the safety structural puzzle. The structure of the Safety Director's team depends upon company policy and the objectives, and therefore can be simple or complex.

An organizational safety structure takes effort to construct. Policy drives its form; corporate objectives mandate the final safety organizational chart, or cause a new structure as objectives evolve. A good way to start is to answer a series of questions that define the organization and its mission. These questions will shape the organizational safety structure.

- How big is the organization?
- What are the major hazards?
- How complicated are the activities?
- What is the nature of risk in the organizational activities?
- What is the organization's tolerance to risk (determined by the CEO)?
- What are the safety resources (determined by the CEO)?

A small company might have one industrial safety professional who oversees the operation, and reports to the president, or perhaps the safety officer also has another job, where the safety role is part-time. Large organizations will have a team of safety professionals, who are educated in safety and skilled with years of experience. Safety managers are key to orchestrating the safety effort, but in reality, they are just part of the whole team. An organization may have an organizational chart displayed in a book showing lines connecting each department to safety. While nice to look at, it does not get the job done. To ensure it is *safe enough*, an organization needs a 'working' structure that gets every member involved.

Whole-team Safety Structure

- CEO establishes policy and goals
  - o Requirements/accountability
- Safety professionals administrate the CEO's safety program
  - o Advise senior leaders/oversees risk management efforts
  - o Inspect for compliance and hazards
  - o Ensure all members are adequately trained
- All managers and supervisors assume accountability
  - o Oversee safe operations/ensure policy compliance
  - o Assess and accept risk when appropriate at their level
- Individuals apply safety procedures and are held accountable
  - o Comply with safety policies and procedures
  - o Identify hazards, and report them in a timely manner
  - o Assess real-time risks and inform supervisors
  - o Be safe.

Success is possible with a 'whole-team' approach. The team concept takes effort and resources, and must embrace eight key components

Eight Key Structural Components of a Safety Program

Regardless of the structure defined by an organizational chart, an effective safety program requires a working structure as outlined above, and some key components (Alston, 2003).

1. Communication network. Communication is vital in safety. What good does it do for a person to be aware of a hazard to an organization and not tell others about it? Communication does not have a specific starting point, but comes from individuals at all levels. The CEO may communicate a safety directive, strategy, or concern; a line worker can have an idea for a better safety procedure. Where safety information begins is irrelevant, where it ends up is vitally important. Safety professionals must establish two information conduits:

- *A method to collect safety information.* Anonymous drop boxes, safety idea worksheets, hazard identification forms, organizational meetings, company rewards for safety ideas, and registration on mailing lists for safety publications and circulars.

- *A way to get safety information out to people who need it.* Letters, meetings, company newsletters, training, direct contact, new policies, e-mail 'safety' flashes. Safety information can be either critical or good to know, and must be handled accordingly.
  - o *Critical information*–Provide mandatory briefings or read files (signed off) prior to employees starting their daily work activities, or stop work (if possible) to inform employees of the hazard.
  - o *Good to know information*–Place on bulletin boards or easels, circulate in newsletters, or brief at safety meetings.

Vital points that make safety information distribution work:

- No 'junk' mail–reputation for importance.
- Get it to the people who need it.
- Make it interesting.
- Use an 'eye-magnet' to draw peoples' attention.
- Have a separate safety bulletin board.
- Hold good, interesting safety meetings.

Communication is important for proactive risk avoidance. Communication is also helpful to enhance a participative safety process.

2. Participative approach to set standards and procedures. Ultimately, the CEO establishes policy and sets standards for safety. In reality, however, he or she may give a general safety vision, and merely approve the standards and procedures laid out by the safety professionals. Meetings between the safety staff and mid-level managers and line supervisors to determine standards and procedures are important ways to establish a feeling of 'ownership' among key leaders in the organization. Along with ownership comes 'buy-in' from those who are watching the store to ensure compliance with policies and standards.

3. Devise a method to ensure standards are being met.

'You get what you *inspect*, not what you *expect*.' Donald Rumsfeld, July 2003.

An inspection or audit process will serve the purpose. An annual inspection is a minimal requirement, biannually is better, depending on the type of activity. Often, there are Federal Statutes that direct frequency, content and depth of an audit; safety professionals will know the requirement. Internal teams can conduct the inspections, but it is sometimes beneficial to contract out to professional safety consultants for an unbiased result. Along with annual inspections, spot inspections are very effective to view activities during a normal daily operation. In either case, a list of safety requirements must be established for the inspection.

Inspection checklists are written with the participative approach, where leaders and line supervisors realize certain safety needs, hazards and written procedures. Individuals must also have the opportunity to submit thoughts and ideas. The safety professionals will incorporate governmental requirements, and other safety standards where compliance is critical. An organization can also acquire ready-made checklists from professional safety companies, consultants, or even non-profit organizations (Chapter 10).

The Flight Safety Foundation (FSF) is a non-profit organization that offers safety material to the public. The FSF has produced an 'Airline Management Self-audit' that serves as a great template, and with minor adjustments can serve nearly any organization. The audit covers every aspect of an airline organization, including management, work force, operations training, etc. A review of one aspect of the FSF's management audit reveals insight into a typical self-audit.

Management Structure (Flight Safety Digest, 1996)

❑ Does the company have a formal, written statement of corporate safety policies and objectives?
❑ Are these adequately disseminated throughout the company? Is there visible senior management support for these safety policies?
❑ Does the company have a flight safety department or a designated flight safety officer?
❑ Is the department or safety officer effective?
❑ Does the department/safety officer report directly to senior corporate management, to officers or the board of directors?
❑ Does the company support periodic publication of a safety report or newsletter?
❑ Does the company distribute safety reports and newsletter from other sources?
❑ Is there a formal system for regular communication of safety information between management and employees?
❑ Are there periodic company-wide safety meetings?
❑ Does the company actively participate in industry safety activities?
❑ Does the company actively and formally investigate incidents and accidents? Are the results of these investigations disseminated to other managers? To other operating personnel?
❑ Does the company have a confidential, nonpunitive incident-reporting program?

- ❑ Does the company maintain an incident database?
- ❑ Is the incident database routinely analyzed to determine trends?
- ❑ Does the company use outside resources to conduct safety reviews or audits?
- ❑ Does the company actively solicit and encourage input from aircraft manufacturers' product-support groups?

If the answer is 'no' to any of the above, a finding of lost safety potential is noted. The above questions are only for management involvement. The entire audit offered by the FSF has 16 other areas of concern, with many more questions. The end purpose of a safety audit is to find program deficiencies and previously unknown hazards upon which to act.

4. Devise a method to act on identified hazards. When safety people discover a hazard, they do not always have the power to act. Managers who control resources and have accountability have the power to fix safety deficiencies. When safety people sit at corporate-level meetings, deficiencies are brought to the table and decisions made. However, safety decisions that require resources may be put on the back burner for another time. Safety professionals must track all safety problems, deficiencies, and hazards that cannot be fixed immediately, and devise a continual review plan with top management.

5. Response plan for accidents. Regardless how careful and aware an organization is, accidents do happen. The very fact that risk is accepted means that the probability will play out at some point in time, and a loss will occur. The safety team must have a response plan.

Before an Accident Occurs

- Train a pool of investigators ahead of time; have one on call.
- Train other members of a response team.
  - o Trained in first aid and rescue
  - o Trained in fire suppression, or have a special fire fighting team
  - o HAZMAT awareness and handling
- Prepare an investigator response kit.
  - o Phone numbers: Who to call within the organization
  - o Camera
  - o Tape recorder
  - o Notepads and pencils
  - o Checklists on the initial investigation, and on-going investigations.

When an Accident Occurs

- Form the response team; notify authorities.
- Save lives first.
- Determine site safety–avoid injuries to the team.
- Walk through accident site; get a feel for the mishap.
- Take pictures.
- Gather witness testimonies; note transient witnesses and record contact information.
- Preserve evidence for investigators and technical experts.

6. Investigative process. Investigations often begin with the initial response, where trained investigators accompany the response team to preserve evidence. When a response to an accident (whether catastrophic or minor) is activated, or an investigator is simply tasked to do an investigation:

- Take control of the investigation response kit.
- Save lives first, if applicable.
- Determine site safety–avoid injuries to the team.
- Walk through accident site; get a feel for the mishap.
- Take pictures.
- Gather witness testimony, note transient witnesses.
- Preserve and inventory evidence.
- Analyze evidentiary data.
- Determine findings; establish causes.
- Write a final report: make recommendations to prevent a similar accident.

The final step is to record accident data into a company safety database. This valuable information can be analyzed for future accident prevention.

7. Analysis program. Analysis programs are prevention tools. Through analysis, safety managers can examine company data to determine the causes of injuries, lost workdays, and damaged equipment, and devise a plan to prevent similar accidents. An analysis team can also discover industry-wide mishap trends by using national incident/accident data provided by their government. Analysis helps safety professionals transform from accident reporting and investigating to mishap prevention.

8. Safety training program. Awareness is key to prevention. Initial awareness comes from training. Engrained awareness is derived from supervision, experience, and follow-on training. The policy letters discussed earlier in this chapter all referred to safety training as a company priority. CEOs know that poor training programs are viewed as negligent; they also know that deficiencies in training result in accidents that impact productivity. Some types of safety training are:

- *Initial training* prior to new accessions starting work, or current members assigned to a new task. Some topics include proper wear of Personal Protective Equipment (PPE), hazardous aspects of the work activity, hazardous materials handling and storage, proper lifting techniques, company safety policy and procedures, hazard reporting, and risk management.
- *Supervisor training.* When someone is promoted into a supervisory role, they must receive specialized safety training, with emphasis on risk management, human factors, and accountability. They must be trained to a working knowledge on all employee PPE requirements and safety procedures.
- *Annual refresher training* for all organizational members.

The safety structure must have depth throughout the organization, and embrace the whole-team concept. The structure begins at the corporate top, where the CEO is involved in safety and major risk management decisions that affect the organization. All leaders must embrace the safety effort in all activities. Safety professionals must be plugged in at the top, and be represented at mid-level management and on the line. In best cases, safety people can even reach out and touch individual members, physically, verbally or with safety information sources. This type of integration enhances prevention to ensure an organization is *safe enough*.

Part 4. Prevention Methodology

The fourth part of a solid safety program is prevention methodology, which is derived from risk management and system safety principles. The organization's safety structure provides the tools: a communication network that provides incident and hazard reporting, an inspection or audit process, an investigation plan, and an analysis program. The prevention process begins when a hazard is identified, but delivers safety when hazards are dealt with.

The safety managers must devise a system to 'act' on hazards. Some are easy, such as removing a tripping hazard from a walk path, and then counseling the individual who left the hazard there in the first place. Other hazards may require major engineering fixes requiring large corporate resources. One way to assist an organization's prevention methodology is to have a committee or council to hold periodic (monthly) meetings along with key corporate leaders to discuss the known hazards and to evaluate progress and requirements to eliminate or mitigate each. The U.S. Department of Defense refers to such meetings as ESOH Councils, or Environmental, Safety and Occupational Health Councils or Committees. These teams are comprised of safety professionals and key leaders to provide an important avenue to discuss the risks against the organization. They also serve a critical review of available resources to combat the threats. Available solutions are reviewed and brought to senior leaders when resources are required, or the risk is significant. The types of solutions are (Alston, 2003):

Engineering solution. Strives to eliminate the hazard. This is often the most expensive and time-consuming solution. However, it is normally the most effective method of eliminating risk–design it out.

Control solution. When resources or technology are not available to design out a hazard, controls are used to isolate the threat. Roping off a hazardous structure to prevent entry is a control measure until resources are available for engineers to fix and safe-up the structure to eliminate the hazard.

Training solution. If the hazard cannot be removed, people are trained, and warned to be careful. For example, if the organizational activity is driving trucks, the drivers must sometimes drive in hazardous conditions. Some threats cannot be designed out, or controlled; the drivers have to go out into the rain, among the crazy drivers. They are trained to turn their lights on, wear seat belts, drive defensively, and slow down when roads are wet. The problem with this solution is, people make mistakes: the human condition is fallible (Reason, 1990).

Personal protective equipment (PPE) solution. PPE confirms that the hazards are present, and the protective equipment attempts to block the hazard from the individual.

Prevention methodology is not difficult, but requires organizational effort to identify hazards, and then eliminate or control them, or train individuals to avoid the threat, or to protect themselves with PPE. Primarily, prevention takes corporate effort, and often requires corporate resources.

Part 5. Compete for Resources

Most leaders and managers understand the CEO will invest resources in ways to achieve the greatest Return On Investment (ROI). A belief that safety is a profit multiplier will allow resource investment into risk reduction efforts, such as forming a system safety team, or purchasing safety equipment, or installing physical guards around hazards. When comparing resources with risk reduction opportunities, we must seek the *biggest bang for the buck* since resources are limited.

Safety ideas and enhancements must compete with corporate interests and limited resources. It is important to focus safety dollars on safety measures that have the greatest safety ROI.

Example (Hypothetical)

At a fictitious company, Government data show a probability that one person might be killed on the job from slipping on the slick concrete floor and striking his or her head. Other data reveal a strong probability that chemicals stored in a tank outside the office building could explode if struck by lightning, destroying the building and killing many people. You as a leader are aware that your loading dock needs new guardrails; the old rails are unsafe. The organization only has $100,000 available for safety initiatives (decided by the CEO). Here are the options:

- *Option 1*: Paint the entire floor space of the plant with non-slip textured paint: cost $85,000. Hope against chance that lightning does not strike the chemical tank outside. Invest the remaining $15,000 in much needed guardrails on the loading dock.
- *Option 2*: Construct a blast shield between chemical plant and building: cost $120,000 (another corporate interest worth $20,000 must be sacrificed to compensate for the extra safety expense). Forget or postpone consideration of the floor hazard and guardrails, there is no money left for that safety project. Look for financial offsets in the budget to fund the project.
- *Option 3*: Move the chemical storage tank to a safe place: cost $80,000. Repair the guardrails for $15,000, and provide non-slip shoe covers ($5,000) to workers in high-risk areas of the plant that offer similar traction as the non-slip floor paint.

Option 3 is the best choice with greatest ROI. The ROI in this case is preventing fatalities, which also enhances production and organizational survival. If the tank blows, and many people die, the organization may not recover, a consequence that requires moving the tank. Option 2 wastes valuable resources and ignores known hazards. Option 1 also wastes resources by painting the whole floor when slip-on shoe covers will manage the risk. Also, Option 1 ignores the major threat to the organization–a possibility of a fatal explosion.

The lesson: When competing for resources, consider the biggest payback in risk reduction for the investment. Innovative safety measures offer increased ROI in reducing and controlling hazards, which ultimately enhances the bottom line. Essentially, *safety is free*. Without a strong safety program, an organization *will* have costly accidents; with safety, many accidents are prevented. Therefore safety pays for itself.

The safety program can be simple, but gets more complex as the organization grows and its activities become multifarious. Regardless, we cannot overlook the five important parts. The leader of the organization must establish the safety policy; it is the vision and way ahead for safe operations. All leaders are responsible for safety within their span of control, and must ensure that organizational safety objectives are clear, and enhance mission achievement. The overall safety structure 'is' the safety program, and involves a team effort from all organizational members. Eight major safety structural components allow oversight of safe operations, audits the organization, investigates incidents and accidents and analyzes data toward an effective prevention methodology. Lastly, safety programs must compete for resources to enhance the program, and ultimately the bottom line. Leaders should never lose sight of the reality: the winning argument for a solid safety program is the cost of not having one.

Chapter 10

Change: The Way Ahead

'Nothing endures but change.' Heraclitus (540 BC - 480 BC)

'The universe is change; our life is what our thoughts make it.' Antoninus (121 AD - 180 AD)

We have all heard the famous quote, 'The only thing constant is change.' It appears true, as universal players race through the cosmos their relative positions are constantly changing (Comins and Kaufmann, 2002). Earthly actors are also engaged in a multitude of activities...activities require action, action is movement, movement is change. This brings to light an important reality: If change must occur, it can only have two possible outcomes–change for the better, or change for the worse.

Your organization will change; will it change for the better? It can only improve or get worse. Human intervention in naturally occurring change can nudge change in a positive fashion. Healthy, productive organizations do not resist change, but forge it in a way to enhance the organization. This is true for product type, market share, mission, corporate processes, structure and every aspect of the operation. It is also true for safety processes and risk reduction methods. *An open mind for positive change is the requirement, not change for change sake.* Leaders can intellectually move forward with progress, but it all boils down to choice; our choices today to safe-up our organizations will be our future in the months and years ahead, and determine *how safe is safe enough.* The tools to enhance organizational safety are available; leaders must choose to use them.

Beginning a Positive Change Process

Change will occur, and positive change is possible. Where should a leader begin to make positive change? A good place to start is to find out where the organization is 'now' regarding safety. A leader can decide on change by reviewing the following:

1. Review the human condition of the organization (stress & perceptions).
2. Examine the corporate policy on safety.
3. Review organizational safety goals.

4. Examine organizational structure to ensure a 'whole-team' approach to safety.
5. Determine if the organization has the right people to manage the safety effort.
6. Access commercial and public safety information and program resources to see what is available.

1. Review the Human Condition of the Organization

As stated in Chapter Seven, Reason (1990) makes the assertion:

- Fallibility is part of the human condition
- We cannot change the human condition
- We 'can' change the conditions under which people work.

I would add a footnote to Reason's second point. While we cannot change our fallible condition, we can change certain behaviors. Humans can learn and change behavior, which changes performance. However, his point is true, we are human and have certain limitations and physiological frailties. There is a tool available that allows a CEO and other leaders to determine if the human condition is *safe enough* within the organization by assessing stress levels and perceptions: *Organizational Safety Assessments.*

The Organizational Safety Assessment (OSA)

A professional team that specializes in assessing individual and group stress levels and perceptions conducts an OSA. These teams are available in the private and public sectors; a typical OSA program is one that is used in the U.S. Air Force. Its purpose is mishap prevention through hazard identification and risk mitigation (AFI 91-202, 2002), and is intended to aid the Commander (or CEO) in risk assessment, decision-making and risk mitigation. It is important to realize, an OSA is a prevention tool, not a crisis response tool following a catastrophic event or concurrent with another major event. Careful steps measure the human condition, and report to the top leader, and begin with OSA team selection.

An OSA team is a structured process, yet adapts to fit the needs and circumstances of the organization. A trained, certified psychologist heads the team, and possesses a state license in psychology. Credibility is important–that is why the license is essential. The team members can vary, depending on the desired outcomes, but generally include expertise in the following areas: human factors, risk management, safety, operators and maintainers in the organizational schemes from clerical work to supply to operations. The team members must have a general knowledge of the organization's purpose and operation to best assess the organization's people.

An OSA assesses and quantifies personal stress levels and perceptions through collection of both objective and subjective data from organizational members. *Objective* data are acquired prior to the OSA Team's arrival on site using validated survey measurement tools, which quantify members' stresses, strains, coping abilities, and perceptions of safety practices and issues. *Subjective* data are gathered during the visit through personal interviews. The OSA process culminates with a briefing presented to the organizational leader on the combined objective and subjective findings. The briefing provides the leadership with member perceptions and attitudes, strengths and weaknesses, and includes recommendations that may assist leaders and supervisors in reducing accident potential (AFI 91-202, 2002).

An OSA is an excellent decision tool to determine where change is needed. Organizational leaders, for a variety of reasons, often fail to see or understand problems, frustrations, and dynamics that exist in their own work force. Even the best leaders, good people, sometimes do not get the word if they have mid-level managers who 'sanitize' or filter upward information and are bottlenecks, or are afraid to tell the boss bad news. Leaders should understand, that by reason of their position, they do not always know what occurs in their organization. Therefore, they must actively seek cultural and climatic information if they want to stay informed. The OSA helps illustrate where deficits exist and why. It helps make 'unknown' factors 'known' and gives an idea of the severity of issues. Subsequent looks can measure change; helping to determine the effectiveness of steps taken to improve safety practices. In the end, an *Organizational Safety Assessment* helps determine the state of safety within the organization, and leads to accident prevention. Once this is accomplished, a policy review is required to determine if changes are in order.

2. Examine the Corporate Policy on Safety

Policy was discussed at length in Chapter 9. However, a periodic review of the basic policy elements is important, especially following an OSA. Policy changes along with other organizational areas help determine the need for change. While each organization tailors their policy for specific ends, basic policy elements include:

- Embrace the 'whole-team' concept for safety and risk reduction.
- Point out the import role each member plays in safety, which ultimately leads to mission accomplishment.

- Define individual and group responsibilities toward safe operations; clearly state corporate expectations.
- Establish the organization's determination for accountability.
- Direct compliance with all relevant Environmental, Safety and Health (ESH) laws and regulations.
- Integrate ESH management decisions into business decisions.
- Establish a requirement to create and maintain a safe working environment and prevent workplace injuries.
- Direct all members to report, assess, and manage EHS risks.
- CEO must sign the policy.

Organizational policy is important, and compliance is critical to retain its relevance. Senior leaders must be open to change as the world changes around them, but change in smart ways that contribute to risk reduction. As policy is reviewed for possible change to ensure an organization is *safe enough*, leaders should also review organizational safety goals.

3. Review Organizational Safety Goals

Safety goals follow policy, and support the mission. They must also be measurable. As in policy, organizations tailor goals to achieve their mission, but essential items to all organizations are:

- Train all members in the safest methods of job performance (initial training and refresher training; individuals and supervisors).
- Train members in, and encourage, reporting hazards, such as:
  o Failure to use personal protective equipment (shoes, safety glasses, hearing protection, etc.).
  o Unlabeled chemicals.
  o Exposed or unsafe wiring.
  o Blocked fire exits.
  o Unsafe driving or failure to wear seat belts.
  o Working in high places without fall protection.
  o Working beneath heavy, suspended loads.
  o Working on electrical or powered equipment without following appropriate lockout, tag-out procedures.
  o Failure to comply with health, safety or environmental regulations and procedures.
  o Environmental, health or safety hazards or accidents.
- Train all members in use of risk assessment tools (Chapters 5 & 8).
- Recognize employee achievements in accident prevention (Chapters 2 & 9).

- Involve employees in the development of prevention procedures and safety processes (whole-team approach) through meetings or suggestion forms (Chapter 9).
- Protect resources (people, equipment & money) by using risk management tools (sometimes difficult to measure) (Chapters 5 & 6).

Safety goals enhance the bottom line and serve as a profit multiplier. If measured properly, goal achievement serves as a safety indicator for the organization. Goal achievement also helps corporate image in litigation situations, displaying the goodwill and positive efforts toward the environment, safety and health. At the end of the day, goals are achieved when all organizational players support the safety effort.

4. Examine Organizational Structure to Ensure a 'Whole-team' Approach to Safety

Organizations must not leave any member out of the safety process. If someone is left out the organization is not *safe enough*, change that situation. Individual members must also sign up to ensuing changes for improvement. To truly have a team effort, all members must play in the risk game. Here are the safety team's roles (Chapter 9):

- CEO approves policy and goals.
  - o Lays out the requirements.
  - o Holds members accountable.
- Safety professionals administrate the CEO's safety program.
  - o Advises senior leaders.
  - o Oversees risk management efforts, and the hazard identification and reporting process.
  - o Performs inspections for compliance and hazards.
  - o Ensures awareness is heightened.
  - o Ensures all members are adequately trained.
- All leaders, managers and supervisors assume accountability.
  - o Oversees safe operations.
  - o Ensures safety policy and statute compliance.
  - o Assesses and accepts risk when appropriate at their level.
  - o Call 'knock it off' when risks are too high.
- Individuals apply safety procedures and are held accountable.
  - o Comply with safety policies and procedures.
  - o Identify hazards, and report them in a timely manner.
  - o Assess real-time risks and inform supervisors.
  - o Be safe.

An organization's safety structure most likely has a wiring diagram shown in a book. However, the effectiveness of the structure is found in the players. It begins with the CEO's policy, and ends with individuals performing tasks in safe ways. Safety managers are key in the process, because they advise the CEO and other leaders, and shepherd the safety program. It is important to have the right people as safety managers.

5. Determine if the Organization Has the Right People to Manage the Safety Effort

Safety professionals are key to the organization's safety program. They know the laws, have studied the art of risk reduction, and often know where hazards abound. They enhance the effort at every step, and must have:

- Experience as a safety professional (or be under the supervision of one).
- A Certified Safety Professional (CSP) Certificate (or be working on one).
- Energy, and a passion for the job.
- Respect from the organization.

The fourth bullet is important as the safety manager administers the CEO's program. If the safety staff has the first three bullets, then they will earn the fourth bullet. Safety managers are in tune with the organizational needs. As changes occur, they know where to go to find safety resources and guidance from outside the corporate body if necessary.

6. Access Commercial and Public Safety Information and Program Resources to See what is Available.

Safety managers should explore new ideas from the field, and determine industry breakthroughs in safety. Particularly as change occurs within an organization, leaders along with safety managers must examine new approaches to the safety process.

Whether a minor adjustment is needed in a safety process, or if starting a safety program from scratch, an organization can review available avenues for help. Commercial products are offered 'on-line' that help organizations in every aspect of a safety program: a general safety program structure (editable for specific needs), training guides, checklists, safety procedures, and more.

One example of a ready-made safety and health program acquired commercially on the Internet is found at www.oshasafety.com/. Safety items available from the product are listed in Figure 10.1.

Back and Lifting Safety	Electrical Safety	Forklift and Motorized Truck	Landscape and Grounds Maintenance	Respiratory Protection Program
Blood borne Pathogens	Ergonomics	General Shop and Work Area	Machinery & Machine Guarding	Roof Labor Safety
Carpentry and Lumber Handling	Excavation and Trenching	Hearing Conservation Program	Motor Vehicle Safety	Safety & Health Signs and Tags
Chemical Safety- HAZCOM	Fall Protection	Heating Systems and Boiler	Office Safety	Scaffold Safety
Concrete and Masonry Construction	Fire Prevention	Housekeeping and Material Storage	Painting Operations	Temporary and Contract Workers
Confined Space	Flammable Liquids	Laboratory Health and Safety	Personal Protective Equipment	Tool Box Talks/ Work Group Safety Meetings
Corridors and Outside Walkways	Fleet Motor Vehicle Safety Program	Laboratory HAZCOM Chemical Safety Plan	Plumbing Operations	Welding, Hot Work, and Metal Fabrication
Cranes and Hoists	Food Service	Ladder Safety	Refrigeration and Air Conditioning	Workplace Hazards and Workplace Violence

Figure 10.1 Safety materials found at www.oshasafety.com/

Another example of a commercial Internet site for commercial safety products is at www.safetynext.com/index.cfm?source=MKD&effort=36, which offers safety tools that help in the following areas:

Checklists	Multimedia	Safety quizzes
Forms	Policies	Safety talks
Handouts	Posters/clip art	Training meetings
Letters	Slide presentations	Written programs

The site also offers library topics, such as:

Audits	General safety	Noise
Confined spaces	Hazard communication	OSHA Regulations
Electrical	Hazardous materials	PPE
Emergencies	Health care	Record keeping
Ergonomics	Inspections/violations	Committees
Fire	Job hazard	Slips and falls
First aid	Lockout/tag out	Training

There is much information available at reasonable prices for interested managers. Other excellent conduits to safety guidance and materials are found in government websites worldwide. In the U.S., an excellent government site for safety is from the Department of labor, Occupational Safety and Health Administration (OSHA). Their website, www.osha.gov, is a goldmine of safety information and strategies. The site offers a litany of materials and products to help organizations in safety efficiency and federal compliance.

Interesting items offered at the OSHA site are *eTools*. eTools are 'stand-alone', interactive, Web-based training tools on occupational safety and health topics. They are highly illustrated and utilize graphical menus. Some also use expert system modules, which enable the user to answer questions, and receive reliable advice on how OSHA regulations apply to their work site (OSHA, 2002). Other product topics for compliance assistance are:

- Laws and regulations
- Consultation
- Grants
- Posters
- Record keeping
- Training.

The OSHA site offers many opportunities for serious organizational leaders and safety professionals. Whether entering voluntary protection programs

(VPP) or forming strategic partnerships, this site supports building an initial safety program, or assists current program seeking change.

The OSHA site offers good advice and tools for any nation. However, other nations' governments also provide excellent assistance for safety programs. One such website is from the government of the United Kingdom, www.hse.gov.uk/. Its purpose it to promote the message that: 'Good Health and Safety is Good Business'. This site provides information sources, safety leaflets, research and statistics (Health and Safety Executives, U.K., 2002). Another outstanding Internet tool is a 'cooperative' effort from Canada and the European Union for 'Workplace Safety and Health' at www.eu-ccohs.org/sitemap/ (EU-CCOHS, 2002). The European Union offers useful suggestions and user-friendly tools to enhance safety. The Canadian Centre for Occupational Health and Safety (CCOHS) hosts 'Health and Safety Canada', an e-mail-based discussion group for people with an interest in health and safety in a Canadian context. CCOHS also provides a list of approximately 250 other health and safety discussion groups in its 'Health and Safety Internet Directory'. Worldwide there are useful Internet sites to help the interested safety professional to find the latest tools and techniques for successful safety programs.

A Review

Let's review. Leaders determine the needs for positive change by reviewing where the organization is 'now'. The leader must answer the question, 'Where are we now in our safety efforts?' Follow that question with, 'Where should we be?' The following steps are considered:

1. Review the human condition to assess stress and perceptions.
2. Examine the corporate policy on safety.
3. Review organizational safety goals.
4. Examine organization for a 'whole-team' approach.
5. Get the right people to manage the safety effort.
6. Access commercial and public safety information.

These six steps help determine if change is indeed needed in the safety program, and if so, conclude the kind and level of changes. Determining change is a key role for organizational leaders when answering the question, *'How safe is safe enough?'* In the process, leaders must always keep their focus on one safety theme: attack risk, and win.

Chapter 11

How Safe is Safe Enough?
The Answer

The terrorist attacks of 11 September 2001 defined a moment for America, and the world. An equally defining moment was the Space Shuttle Columbia break-up over Texas on 1 February 2003. Both catastrophic events cause us to wonder if we are *safe enough*. Forever changed, we now live in an era where we must confront security and safety risks before the result–death and destruction. With awareness, focus, and effort, we can manage the risks to our organizations and ourselves. Our opportunity, as individuals and organizations: change with our nation, change with the world. We realize we must coexist with risks of every kind, but now have awareness that we can intervene in the natural events of probability and severity. Leaders must consider they are not alone and unarmed when confronting organizational risks, and have tools available to answer the question, 'How safe is safe enough?'

Leaders at all levels must address the question of safety. They control resources and define the mission; they bear ultimate accountability for the performance of an organization. Safety remains a key factor in organizational performance, and can either enhance or detract from the bottom line depending on its level of attention. Leaders shoulder the distinctly important task of weighing safety training and safety technology with limited resources. They must also consider safety's impact on mission elements, and seek out and eliminate hazards in operations as well as risks to the community. During their thought processes, leaders must decide what level of risk is acceptable, and if probabilities play out in the worst way, determine if the organization can survive the result.

Necessary Risk

Some risks are necessary since risk comes with every activity, but how much is acceptable? How safe is safe enough? The answer is this: *an organization is safe enough when the leader seeks out modern safety processes, and makes the effort to identify every possible hazard, and then strives to eliminate, control, or reduce the associated risks through training, procedures, and technology to the point that operations do not accept unnecessary risks.*

The above answer begs the next question, 'Which risks are necessary?' Necessary risks are *those irresolvable, residual risks found in the operational activities that benefit the organization and accomplish the mission.* For example, a trucking company faces many road hazards, yet must put trucks on the road to make a profit. By releasing resources for sound safety training and safety devices (technology) on the trucks, leaders reduce and control risks. However, some residual acceptable risks remain as their drivers face damaged roads, inclement weather, and crazy drivers on the highways. In all cases, leaders need to focus on the basic rule of safety: *The benefit of any activity must outweigh the risk* and contribute to mission accomplishment, otherwise risk is too high for the payback of the event, and is unnecessary.

How Leaders Win the Risk Game

In our quest for safety, we need to provide the basics, knowing we will not cover everything or be perfect. However, if leaders acquire the right people for the organization, give them the right training and the right equipment, and establish sound safety processes, the organization is generally 'safe enough'. To best prepare themselves to determine if their organizations are *safe enough*, leaders need to familiarize themselves with concepts and tools that reduce risk. The preceding chapters in this book have laid out nine elements that leaders must know to shape their safety culture: rules of the road to win the *risk game*.

1. Leaders must lead the risk game. Leaders show the way. Organizational leaders have two responsibilities in safety:

    *   A *fiscal* responsibility to enhance the bottom-line through sound safety practices and a solid risk management process.
    *   A *moral* responsibility to organizational members, their families, stakeholders, and society in general to protect life, guard corporate assets, and preserve the environment.

    Another important concept is that leaders achieve *buy-in* from their organizational members with five tasks:

    *   Hold people accountable for negligence and willful disregard for safety.
    *   Provide incentives for positive safety performance.
    *   Mentor corporate members on safety and risk management.
    *   Create a participative safety process for all members.
    *   Be a champion of safety.

2. Know the costs of losing the risk game.

    - The force behind a safety program is the cost of not having one.
    - There are tangible and intangible costs: both impact production.
    - Many stakeholders are at risk.
    - Some costs, as in Chernobyl, affect society as a whole.
    - Preserving assets multiplies profits.

3. Comprehend universal probabilities and the effects of human intervention.

    - Random events both build and destroy worlds.
    - Humans perceive and interact with the world; we record events and pass them along.
    - We can create opportunities to intercede in natural events.
    - Human intervention alters probabilities and severities.

4. Understand basic principles of risk management.

    - Six-step process to control risks.
    - Risk management matrix models help to choose risk reduction options.
    - 'Real-time' operational risk assessment worksheets evaluate current levels of risk.
    - Personal morning risk assessment checklists help individuals to know their personal risk level.

5. Understand the basics and know the value of the system safety process.

    - Define a hazard.
    - Define the system.
    - Identify hazards.
    - Assess the risks.
    - Eliminate or reduce risks.

6. Be familiar with elements of organizational risk.

    - The 'human factor' stands in the way of reaching the goal of 'zero' mishaps. Individuals are fallible.
    - Individuals make up organizations.
    - There are three levels of risk acceptance: strategic, operational, and individual.

- Individual perceptions impact on risk acceptance.
- Environmental risks outside the organization pose risk to its survival.
- Internal factors present risk that must be managed in the following organizational areas: work ethic, structure, behavior, culture, and climate.

7. Appreciate the value of personal risk management.

- Individuals need to know their responsibilities to themselves and others to avoid risk.
- Seek out risky personal behavior, and control it.
- Use a Personal Risk Management Guide.
- Utilize a simplified version of risk management.
- Embrace six personal risk reduction rules.

8. Get involved in the organization's safety program.

- Craft a safety policy letter/statement from the CEO, or organizational leader.
- Set clear safety objectives to help achieve the mission.
- Develop an organizational safety structure in depth, with qualified personnel, and required key components.
- Devise an accident prevention methodology.
- Help safety compete for resources.

9. Be open to positive change.

- Review the human condition to assess stress and perceptions.
- Examine the corporate policy on safety.
- Review organizational safety goals.
- Ensure the organization has a 'whole-team' approach.
- Get the right people to manage the safety effort.
- Access commercial and public safety information sources for modern safety tools and techniques.

These nine elements are designed to fight and win against risk. CEOs and managers at every level can apply these winning tools and enjoy the ensuing success.

Added to the nine elements listed above are the following seven 'truths' of safety along the road to zero mishaps:

1. All mishaps are avoidable.
2. Achieving 'zero' accidents is possible (though often blocked by the human factor; the fallible human condition).
3. Risk comes with all activity: the benefit of an activity must outweigh the associated risk.
4. Safety is integral to operational success.
5. While everyone is responsible for safety, leaders have ultimate responsibility.
6. Without accountability, no one is responsible for safety.
7. An organization is only as safe as the leader allows it to be.

If all accidents are avoidable, then the possibility of achieving 'zero' accidents exists. Leaders must strive for that prize–zero mishaps. However, we must be aware that every activity bears a certain level of risk, and over time probabilities may play out (Chapter 4). In the complexities of organizational activities (which all bear risk), and considering the fallible human condition, it is unlikely we will reach and sustain a level of zero accidents in our lifetimes, even though zero is the ideal goal. It is in this light, and at this time in our human evolution, that leaders must strive for an optimum 'safe enough' organization.

Leaders and managers can fight and win the risk game to be *safe enough*, but it takes a concerted effort. Harnessing the human condition remains the overarching challenge; addressing the human factor is central to any accident prevention effort. Chapters 7 and 10 discuss James Reason's assertion that the human condition is fallible, and that we cannot change that fact, which renders us the largest roadblock on the road to zero. There are four roadblocks of concern:

1. Acts of God
2. Unavailable resources
3. Unavailable technology
4. The Human Factor.

Of these, the human factor is the greatest obstacle to reaching the elusive 'zero'. Paradoxically, humans are key to winning the risk game, because we perceive and interact with our environment, and we record and therefore pass along knowledge, unlike any other known species (Norman, 1990). We can intervene in our environments through commitment of resources and technology (when available) to protect us from ourselves. We as humans are subject to human error and psychological and physiological frailties such as fatigue, perceptions, stress, complacency and distraction. While we can improve our performance with sound training and mentoring, we cannot change our basic condition. We can, however, change the conditions in which we work to protect ourselves from our less than perfect states. Physical safeguards, personal protective equipment, systems safety, safety procedures, and careful supervision help 'safe-up' our work environments, and enhance our journey toward zero mishaps. We can choose, we can change, and we can overcome what nature throws our way. That is our unique human challenge, and our heritage.

Unsafe practices and unnecessary risk acceptance, both on the job and off the job, harm an organization. Members who are injured or killed while camping or hiking present as great a loss as those at work on the line. The tools outlined in this book apply to all organizations, whether pursuing profit or not, whether accomplishing job tasks or having fun. All organizational leaders will benefit from applying the nine elements of risk reduction and maintaining an awareness of hazards and possible interventions. With effort and sincerity, all have the ability to win the risk game on the road to zero mishaps, and do so by continually answering the question, '*How safe is safe enough?*'

Bibliography

AFI 90-901 (Air Force Instruction) (2000), *Operational Risk Management (ORM)* Command Policy, U.S. Air Force Publications Office.

AFI 91-202 (Air Force Instruction) (2002), *The US Air Force Mishap Prevention Program*, U.S. Air Force Publications Office.

Air Combat Command (2002), *Personal Risk Management (PRM), Summer Guide, Risk Assessments*, Air Combat Command Safety Office.

Alston, G. (2003), 'Lecture Notes' (Compilation of 15 years of lecture notes as an Assistant Adjunct Professor for Embry-Riddle Aeronautical University, 1988-2003. Some notes are from unknown sources).

Bahr, N. J. (1997), *System Safety Engineering and Risk Assessment: A Practical Approach*, Taylor & Francis, Philadelphia.

Barnett, A. (2002), 'Aviation Safety and Business School Statistics', Massachusetts Institute of Technology, a briefing on the Web Page, http://roger.babson.edu/rao/abtalk.ppt.

BCSP Board of Certified Safety Professionals (2002), 208 Burwash Avenue, Savoy, IL 61874, Web Page, http://www.bcsp.org/assoc.html.

Bernstein, P. L. (1996), *Against the Gods: The Remarkable Story of Risk*, John Wiley & Sons, Inc., New York.

Broome, J. (1978), 'Trying to Value a Life', *Journal of Public Economics*, 9:91-100.

Chaplin, J. P. and Krawiec, T. S. (1960), *Systems and Theories of Psychology*, Holt, Rinehart & Winston, New York.

Chernobyl (2002), 'Understanding Some of the True Costs of Nuclear Technology' (Unknown Author), Web Page, http://www.ratical.org/radiation/Chernobyl/.

Comins, N. F. and Kaufmann, W. J. (2002), *Discovering the Universe*, 6th edn, W. H. Freeman and Company, New York.

David, F. L. (1962), *Games, Gods, and Gambling*, Hafner Publishing Company, New York.

Deal, T. E. and Kennedy, A. A. (1982), *Corporate Cultures*, Addison Wesley, Reading, Massachusetts.

Department of the Army, Seattle District, Corps of Engineers (2002), 'Policy Letter', Web Page, http://www.nws.usace.army.mil/SafetyOffice/policy/NWS.

Dillinger, T. G. (1994), Psy. D., 'The fighter pilot in distress'. Unpublished doctoral dissertation: The Chicago School of Professional Psychology.

EU-CCOHS (2002), 'Canada & European Union Cooperation on Workplace Safety and Health', Web Page, http://www.eu-ccohs.org/sitemap/.

Evans, L. (1991), *Traffic Safety and the Driver*, Van Nostrand Reinhold, New York.

Ferry, T. S. (1981), *Modern Accident Investigation and Analysis*, John Wiley & Sons, New York.

Flight Safety Digest, Vol. 13, No. 12, p. 1, Dollars and Sense of Risk Management and Airline Safety, Flight Safety Foundation, Washington D.C.

Flight Safety Digest, Vol. 15, No. 11, p. 1, Aviation Safety: Airline Management Self-audit, Flight Safety Foundation, Washington D.C.

General Electric Company (2002, 'Safety Policy', Web Page, http:// www.ge.com /commitment/social/integrity/ehs.htm.

Ginley, J. (2002), 'An Internal Audit Perspective', a briefing, Public Works and Government Services Canada, Web Page, http://www.spin.org/Ginley/sld001.html.

Green, M. and Senders, J., (1997), 'Human Error in Road Accidents', ERGO/GERO Human Factors Science, Web Page, http://www.ergogero.com/pages /roadaccidents.html.

Health and Safety Executive, U.K. (2002), 'Good Health and Safety is Good Business', Web Page, http://www.hse.gov.uk/.

Health and Safety Services UK Ltd. (2002), 'The Hidden Costs of Accidents', Web Page, http://www.healthandsafetyservices.biz/.

Krause, S. S., (1996), *Aircraft Safety: Accident Investigations, Analysis, & Applications*, McGraw-Hill Companies, Inc., New York.

Kunreuther, H. and Slovic, P. (1996), *Challenges in Risk Assessment and Risk Management*, Sage Publications, Inc., Thousand Oaks.

Lockheed Martin Aeronautics Company (2002), 'Safety Policy' Web Page, http://www.lmaeronautics.com/initiatives/esh/index.html.

Mearns, K., Whitaker, S., Flin, R., Gordon, R., and O'Connor, P. (2000), *Factoring the Human into Safety: Translating Research Into Practice*, Rep. No. HSE OTO 2000 061.

Nelson, D. L. and Quick, J. C. (2000), *Organizational Behavior: Foundations, Realities and Challenges*, 3rd edn, South-Western College Publishing.

Norman, D. (1990), *Design of Everyday Things*, Bantam & Doubleday, New York.

OSHA (U.S. Occupational Safety and Health Administration) (2002), Web Page, http:/www.osha.gov.

Pearce, J. A. and Robinson, R. B. (2000), *Strategic Management: Formulation, Implementation, and Control*, 7th edn, McGraw-Hill Higher Education, Boston.

Reason, J. (1990), *Human Error*, Cambridge University Press, Cambridge.

Report of the Presidential Commission on the Space Shuttle Challenger Accident (1986), Presidential Commission on the Space Shuttle Challenger, Washington, D.C.

Roland, H. E. and Moriarty, B. (1990), *System Safety Engineering and Management*, revised edn, John Wiley, New York.

Schlumberger Technology (2002), 'Safety Objectives', Web Page, www1.slb.com/qhse/safe/.

Slovic, P., Fischhoff, B., Lichtenstein, S. (1979), 'Rating the Risks', *Environment*, vol. 21, no. 3.

Weigman, D. A., Zhang, H., Thaden, T., Sharma, G., and Mitchell, A. (2002), 'A Synthesis of Safety Culture and Safety Climate Research', Aviation Research Lab, Institute of Aviation: University of Illinois at Urbana-Champaign.

Index